General Editor
Jeff Daniels

COLLINS
London and Glasgow

First published 1986
Wm Collins Sons and Co Ltd

©Chevprime Ltd

Produced by Chevprime Ltd
147 Cleveland Street
London W1
London NW1 3HG

ISBN 0 00 458838 X

Artwork by Maltings Partnership,
Mike Badrocke and Pilot Press

Filmset by Tradespools Ltd, Frome

Printed by New Interlitho, Italy

Contents

Introduction

The final few months of World War II saw the introduction of the first jet combat aircraft, and wrought a change in military aircraft design at least as significant as the switch from the biplane to the monoplane during the early 1930s.

The only Allied jet to see active service was the Gloster Meteor; the main German type involved in combat was the Messerschmitt 262 although there were occasional reported encounters with the Heinkel 162 light fighter and the Arado 234 light bomber. The war ended before two other Allied types, Britain's de Havilland Vampire and Lockheed's Shooting Star, reached front-line squadrons. The real surprise, though, came when Allied forces had the chance to inspect the remains of German factories after hostilities were over. They found evidence of a dozen or more very advanced designs in the prototype stage, including projects such as a four-jet medium bomber with forward-swept wings, and a fighter with variable-geometry swept wings.

There was no chance that the end of World War II would see the near-total disbandment of the Allied air forces and the cessation of practically all advanced aeronautical research, as had happened in 1919. Even as the British and American investigators gathered evidence and knowledge in factories in that part of Germany which they occupied, it became clear that the Russians were doing the same – and Russian-occupied Germany happened to include

Peenemunde which was not only the launch pad for V-weapon tests, but also a major flight-test centre for advanced prototype aircraft. It was equally clear that the erstwhile allies, America and Britain on the one hand and Soviet Russia on the other, had little enough in common (once Germany had been defeated) that they would regard each other with huge distrust. That in turn meant that each side kept substantial military forces after the end of the war, and that research into advanced military aircraft continued.

Both sides learned a great deal from German experience in both jet engines and advanced airframes. On the airframe side the most valuable German research had been into swept and delta wings which were the key to high-speed flight up to and beyond Mach 1 – the speed of sound. Where engines were concerned, the real interest lay in the Germans' work on axial-flow types (in which the air flows directly from front to rear of the engine and both the compressor and the turbine consist of dozens of small aerofoil blades) while the early British engines – licensed in turn to the Americans – were centrifugal-flow (that is, the compressor was a solid disc with vanes that flung the air outwards as it was compressed). The British engines were simple, tough and reliable but it was the axial-flow type which held out the real promise of future development.

While designers everywhere studied the problems of swept wings, more and more straight-winged jet

5

fighters entered production. In Britain, the Venom succeeded the Vampire and the Attacker and Sea-hawk were developed for naval use. In the USA, the Shooting Star was joined by the F-84 Thunderjet. The French built the Ouragan and behind the Iron Curtain the first Russian types, notably the MiG-9, were also flying.

The first major swept-wing fighter type to fly in the West was North American's F-86 Sabre which was actually an adaptation of a straight-winged type (The FJ Fury) which had been developed as a carrier-borne machine. It was just as well the Sabre was a good design and that it was quickly put into production, because when the Korean War began in 1950 the Americans encountered, totally unexpectedly, a swept-wing Soviet fighter – the MiG-15. But for the Sabre's ability to outfly the MiG, life would have been much more difficult for the ground-attack aircraft which had to strike at North Korean and Chinese forces.

At this time, the RAF had not a single swept-wing aircraft in service. There was acute embarrassment, indeed, that a country as small as Sweden had managed to get a highly capable home-built swept-wing fighter – the Saab 29 – into squadron service while British defences still depended on later marks of Vampire and Meteor. Such was the gap that Britain was forced to buy Sabres from the USA until its own Hawker Hunter was in full production. The French, meanwhile, had embarked on their own very successful series of Mystère fighters (initially developed from the straight winged Ouragan).

Where bombers were concerned, the picture was more confused. In the years immediately after the war it was assumed that jet engines were strictly for fighters, because their fuel consumption meant no bomber could be built with adequate range. The USA went ahead and built a huge piston-engined bomber, the B-36, with a wing span far larger than that of a modern Jumbo. The British, for all their slow development of supersonic fighters, took the view that jet bombers could be made to work and produced first the Canberra medium bomber and then the trio of heavy 'V-bombers' – Valiant, Victor and Vulcan. All except the Valiant were destined to serve the RAF for nearly 30 years. The Americans took the point and themselves produced two advanced jet bombers, the medium B-47 and the heavy B-52, both built by Boeing. The Russians took far longer to build heavy jet bombers and filled their gap with a unique swept-wing turboprop type, the Tupolev 'Bear'.

While such types were being made ready in the early 1950s, a new generation of truly supersonic types was on the drawing boards. The first American type to fly was the F-100 Super Sabre, from the team which designed the earlier F-86. It was the first of a whole team of 'century-series' fighters which included the big, long-range F-101 Voodoo, the delta-winged F-102 and F-106, the lightweight Lockheed

F-104 with its tiny, straight wings, and the capable F-105 Thunderchief which bore the brunt of many attack operations in Vietnam. At the same time the US Navy ordered its own new supersonic types including one, the McDonnell Phantom, which proved so good that it was eventually ordered by the Air Force as well. Britain's Royal Navy meanwhile took delivery of two fighter types, the Sea Vixen and the Scimitar, together with the very successful Buccaneer light bomber – whose US Navy equivalent is the Grumman A-6 Intruder.

Dassault MIRAGE 111RD
reconnaissance aircraft

The British built only one supersonic fighter, the Lightning, before the Ministry of Defence misguidedly cancelled all further projects on the grounds – in 1958! – that missiles would soon be able to do everything. It was not a decision shared by the Russians, who followed the MiG-15 with a succession of ever more advanced supersonic designs such as the ubiquitous MiG-21 and MiG-25, nor by the French who pressed on successfully with their Mirage series, nor even by the Swedes who built the J-35 Draken and J-37 Viggen.

The truly supersonic bomber proved a more difficult nut to crack. An early American attempt, the B-58 Hustler, entered squadron service but could hardly be called a success: the French did better with their twin-engined Mystère IV. The Russians, it seems suffered a number of disasters before producing the Tu-22 Blinder and the later Backfire. The British might have had an equivalent with the BAC TSR2, but cancelled it.

More recently, the Americans have produced yet another generation of advanced fighters. In effect, the US Air Force and US Navy each has one large, twin-engined type plus one lighter, more agile, single-engined fighter: the F-15 Eagle and F-16 Fighting Falcon, and the F-14 Tomcat and F-18 Hornet respectively. Also in American service is a

derivative of the most successful all-British combat design of recent years, the BAC Harrier VSTOL (Vertical/Short Take Off and Landing) fighter. In recent years, it has otherwise been British policy not to produce its own major military aircraft types, but to work in partnership with other European countries in the belief that this reduces cost. The most successful results of this policy have been the Jaguar light strike aircraft, built together with the French, and the Tornado variable-geometry strike aircraft whose design was shared with the Germans and Italians.

It is interesting that there should be such a strong European interest in aircraft which are more multi-role attack machines than fighters as such. It is a

General Dynamics F-16A Fighting Falcon

Fairchild A-10 Thunderbolt

feeling reflected on the Russian side of the Iron
Curtain where types like the MiG-27 Flogger and
Sukhoi 24 Fencer play similar roles to those of the
Jaguar and Tornado respectively. The trend has been
taken to the extreme that neither Britain nor Germany has a true fighter of modern European design
in service but depend heavily on the McDonnell
Phantom (and in Britain's case, before long, on an
adapted heavy fighter version of the Tornado). The

standard USAF attack type is the variable-geometry
F-111; an interesting addition to the USAF armoury
in Europe is the more recently designed Fairchild A-
10 Thunderbolt, a specialised 'tank-buster' along the
lines of the World War II Soviet *Sturmoviks*.

While most of the interest in military aircraft is
naturally concentrated on fighters and bombers, it
should not be forgotten that other more specialised
types have their own importance. This is especially
true of maritime patrol aircraft which are responsible
above all for countering any enemy submarine men-
ace. Such types may be fairly small and based on

NATO E-3 AWAC

aircraft carriers, like the Fairey Gannet formerly used by the Royal Navy, the French Breguet Alizé and the US Navy's current S-2A Viking; or they may be larger and land-based like the BAe Shackleton (a development of the wartime Lancaster) or the Lockheed Neptune. Today's most advanced and important patrol aircraft are Britain's BAe Nimrod, the French Breguet Atlantic, and the American Lockheed Orion; the Russian equivalent is the Ilyushin-38.

14

Another type of patrol aircraft which has become vital in today's world is the airborne early-warning (AEW) type which acts not only as a long-range target detector but also as an airborne command post for interceptor fighters. Such types are very expensive but, against that, are needed only in small numbers. By far the best known is the E-3 AWACS adaptation of the Boeing 707 airliner. The Grumman Hawkeye carrier-borne AEW machine is a remarkable technical achievement, and the British entry in this class is the BAe Nimrod 3.

Aeritalia G91Y

Country of Origin: Italy
Type: tactical strike and reconnaissance fighter
Accommodation: pilot seated on a Martin-Baker
ejector seat

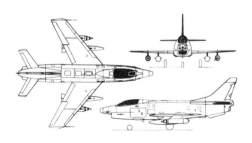

Armament (fixed): two 33mm DEFA 552 cannon
Armament (disposable): up to 1,814kg (4,000lb)
Electronics and operational equipment: communication and navigational equipment
Powerplant and fuel system: two 1,850kg (4,080lb) afterburning thrust General Electric J85-GE-13A
Performance: maximum speed 1,140km/h (708mph); service ceiling 12,500m (41,010ft); range 385km (240 miles)
Weights: empty 3,682kg (8,117lb); normal take-off 7,800kg (17,196lb); maximum take-off 8,700kg (19,180lb)
Dimensions: span 9.01m (29ft 6.5in); length 11.67m (38ft 3.5in); height 4.43m (14ft 6.5in); wing area 18.13m² (195.15sq ft)

Aermacchi M.B.326L

Country of Origin: Italy
Type: basic and advanced trainer

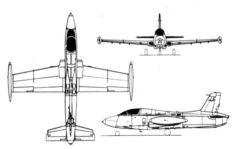

Accommodation: pupil and instructor seated in tandem on Martin-Baker Mk 06A ejector seats
Armament (fixed): none

Armament (disposable): up to a maximum weight of 1,814kg (4,000lb)

Electronics and operational equipment: communication and navigation equipment, plus Doppler radar, SFOM Type 83 fixed or Ferranti LFS 5/102A gyro gunsight, and provision for a bombing computer and laser rangefinder

Powerplant and fuel system: one 1,815kg (4,000lb) thrust Rolls-Royce (Bristol) Viper Mk 362-43

Performance: maximum speed 890km/h (593mph) at 1,525m (5,000ft) clean; service ceiling about 14,325m (47,000ft); range 268km (167 miles)

Weights: empty 2,964kg (6,534lb); normal take-off 4,211kg (9,285lb); maximum take-off 5,897kg (13,000lb)

Dimensions: span 10.85m (35ft 7in) over tiptanks; length 10.673m (35ft .25in); height 3.72m (12ft 2in); wing area 19.35m² (208.3sq ft)

Aérospatiale CM.170 Magister

Country of Origin: France
Type: multi-role trainer and light attack aircraft
Accommodation: pupil and instructor

Armament (fixed): two 7.5 or 7.62mm (.295 or .3in) machine-guns in the nose with 200 rounds per gun

Armament (disposable): underwing rockets, bombs or Nord AS.11 missiles

Electronics and operational equipment: communication and navigation equipment

Powerplant and fuel system: two 400kg (882lb) thrust Turboméca Marboré IIA turbojets

Performance: maximum speed 715km/h (414mph) at 9,145m (30,000ft), and 650km/h (403mph) at sea level; initial climb rate 1,020m (3,345ft) per minute; service ceiling 11,000m (36,090ft); range 925km (575 miles)

Weights: empty 2,150kg (4,740lb); normal take-off 3,100kg (6,834lb); maximum take-off 3,200kg (7,055lb)

Dimensions: span 12.15m (39ft 10in) over tiptanks; length 10.06m (33ft); height 2.80m (9ft 2in); wing area 17.30m² (186.1sq ft)

Aero Vodochody L-39D Albatross

Country of Origin: Czechoslovakia
Type: single-seat light attack aircraft
Accommodation: pilot on ejector seat

Armament (fixed): a pod containing one two-barrel GSh-23 23mm cannon and between 150 and 180 rounds may be attached under the fuselage, used in conjunction with an ASP-3-NMU-39 gunsight
Armament (disposable): a maximum of 1,100kg (2,425lb) of stores

Electronics and operational equipment: communications and navigation equipment

Powerplant and fuel system: one 1,720kg (3,792lb) thrust Walter Titan turbofan (licence-built Ivchenko AI-25-TL) or a 1,900kg (4,188lb) thrust

Performance: maximum speed with stores 630km/h (391mph) at 6,000m (19,685ft); initial climb rate 960m (3,160ft) per minute; service ceiling 9,000m (29,530ft); range 780km (485 miles) with stores, and 1,600km (994 miles) for ferrying

Weights: empty 3,330kg (7,341lb); normal take-off 4,570kg (10,075lb); maximum take-off 5,270kg (11,618lb)

Dimensions: span 9.46m (31ft .5in); length 12.32m (40ft 5in); height 4.72m (15ft 5.5in); wing area 18.80m^2 (202.36sq ft)

Beech T-34C-1 Turbine Mentor

Country of Origin: USA
Type: trainer and light attack aircraft
Accommodation: pupil and instructor

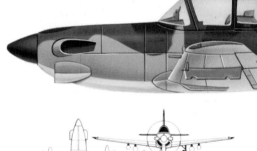

Armament (fixed): none
Armament (disposable): up to a maximum weight of
544kg (1,200lb)

Electronics and operational equipment: communication and navigation equipment, plus a CA-513 fixed-reticle sight

Powerplant and fuel system: one 533kW (715shp) Pratt & Whitney Aircraft of Canada PT6A-25

Performance: maximum speed 386km/h (240mph) at 5,485m (18,000ft); initial climb rate 539m (1,770ft) per minute; service ceiling over 9,145m (30,000ft); range 555km (345 miles)

Weights: empty 1,356kg (2,990lb); normal take-off 1,950kg (4,300lb); maximum take-off 2,494kg (5,500lb)

Dimensions: span 10.16m (33ft 3.9in); length 8.75m (28ft 8.5in); height 2.92m (9ft 7in); wing area 16.69m² (179.6sq ft)

Beriev M-12 Tchaika 'Mail'

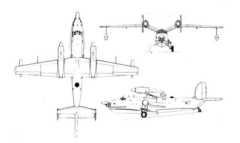

Country of Origin: USSR
Type: maritime reconnaissance amphibian
Accommodation: crew of three or four on the flightdeck, and mission crew of undetermined size
Armament (fixed): none

Armament (disposable): up to a maximum weight of some 5,000kg (11,023lb)

Electronics and operational equipment: communication and navigation equipment, plus search radar in a nose 'thimble', magnetic anomaly detection (MAD) gear in the tail 'sting' and sonobuoys in the fuselage bay, as well as onboard computation and analysis equipment

Powerplant and fuel system: two 3,124kW (4,190ehp) Ivchenko AI-20D turboprops

Performance: maximum speed 608km/h (378mph); cruising speed 320km/h (199mph); initial climb rate 912m (2,990ft) per minute; service ceiling 11,280m (37,000ft); range 4,00km (2,485 miles)

Weights: empty about 21,700kg (47,840lb); maximum take-off 29,450kg (64,925lb)

Dimensions: span 29.71m (97ft 5.75in); length 30.17m (99ft); height 7m (22ft 11.5in); wing area 105m² (1,130.2sq ft)

Boeing B-52G Stratofortress

Country of Origin: USA
Type: long-range strategic bomber
Accommodation: crew of six (pilot, co-pilot, navigator, radar navigator, ECM officer and gunner)

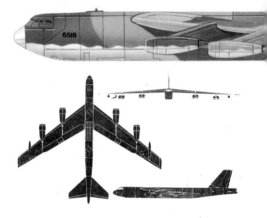

Armament (fixed): four 12.7mm (.5in) M3 machineguns in a remotely controlled rear barbette
Armament (disposable): this comprises 20 Boeing AGM-69A SRAMs (Short-Range Attack Missiles) carried as eight on a rotary launcher in the bomb bay

Electronics and operational equipment: communication and navigation equipment, and a constantly updated electronics suite

Powerplant and fuel system: eight 6,237kg (13,750lb) thrust Pratt & Whitney J57-P-43WB

Performance: maximum speed 957km/h (595mph) or Mach .9 at high altitude, or 676km/h (420mph) at low level; cruising speed 819km/h (509mph) or Mach .77 at high altitude; service ceiling 16,765m (55,000ft); range more than 12,070km (7,500 miles) with maximum internal fuel but without inflight-refuelling

Weights: maximum take-off more than 221,357kg (488,000lb)

Dimensions: span 56.39m (185ft); length 49.05m (160ft 10.9in); height 12.40m (40ft 8in); wing area 371.6m² (4,000sq ft)

Boeing E-3A Sentry

Country of Origin: USA
Type: airborne warning and control system (AWACS)
aircraft

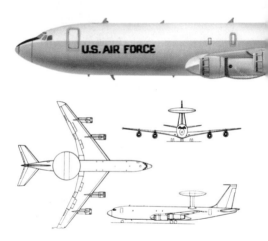

Accommodation: crew of four on the flightdeck, and
a mission crew of thirteen in the cabin
Armament (fixed): none
Armament (disposable): none

Electronics and operational equipment: communication and navigation equipment, plus Westinghouse AN/APY-1 surveillance radar with its antenna measuring about 7.32m (24ft) by 1.52m (5ft) in a rotodome measuring 9.14m (30ft) by 1.83m (6ft) turning at 6rpm when the radar is in use but at .25rpm when it is not, the antenna being backed by IFF/TADIL C antenna (AIL AN/APX-103 system)

Powerplant and fuel system: four 9,526kg (21,000lb) Pratt & Whitney TF33-PW-100/100A

Performance: maximum speed 853km/h (530mph) at high altitude; service ceiling over 8,850m (29,000ft); range 6 hour patrol endurance at a radius of 1,609km (1,000 miles)

Weights: maximum take-off 147,420kg (325,000lb)

Dimensions: span 44.42m (145ft 9in); length 46.61m (152ft 11in); height 12.73m (41ft 9in); wing area 283.35m² (3,050sq ft)

31

Boeing E-4B

Country of Origin: USA
Type: advanced airborne command post (AABNCP)
aircraft
Accommodation: two crews of three or four on the
flightdeck, and about 50 battle staff and communications staff in the cabin

Armament (fixed): none

Armament (disposable): none

Electronics and operational equipment: communications and navigation equipment, and a mass of classified communications, ECM and data-analysis systems provided by Boeing, E-Systems, Electrospace Systems, Collins Radio, RCA, Burroughs and Special Systems Group; the super high frequency (SHF) satellite communication system is distinguishable by the dorsal hump above the upper deck

Powerplant and fuel system: four 23,814kg (52,500lb) thrust General Electric CF6-50E

Performance: range classified, but 12-hour unrefuelled endurance is standard, and a mission endurance of 72 hours is possible

Weights: maximum take-off 361,520kg (797,000lb)

Dimensions: span 59.64m (195ft 8in); length 70.51m (231ft 4in); height 19.33m (63ft 5in); wing area 511m² (5,500sq ft)

British Aerospace Buccaneer S.Mk 2B

Country of Origin: UK
Type: low-level strike aircraft
Accommodation: pilot and systems operator seated in tandem on Martin-Baker ejector seats
Armament (fixed): none
Armament (disposable): this is carried on the revolving door of the weapons bay and on four underwing hardpoints, each rated at 1,361kg (3,000lb), up to a maximum weight of 7,258kg (16,000lb)

Electronics and operational equipment: communication and navigation equipment, plus Decca Doppler radar, central air-data computer, moving-map display, and Ferranti search and fire-control radar with terrain-warning capability and strike-sighting and computing system

Powerplant and fuel system: two 5,035kg (11,100lb) thrust Rolls-Royce RB.168-1A Spey Mk 101 turbofans

Performance: maximum speed 1,038km/h (645mph) or Mach .85 at 61m (200ft); initial climb rate 2,134m (7,000ft) per minute; service ceiling more than 12,190m (40,000ft); range 1,851km (1,150-mile) hi-lo-hi radius with weapons

Weights: empty about 13,608kg (30,000lb); normal take-off 25,402kg (56,000lb); maximum take-off 28,123kg (62,000lb)

Dimensions: span 13.41m (44ft); length 19.33m (63ft 5in); height 4.95m (16ft 3in); wing area 47.82m² (514.7sq ft)

British Aerospace Canberra B(I).Mk 8

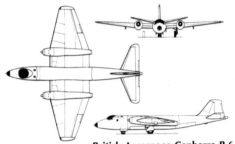

British Aerospace Canberra B.6

Country of Origin: UK
Type: light bomber/intruder
Accommodation: pilot and navigator
Armament (fixed): four Aden 20mm cannon with 500 rounds per gun in a ventral pack carried optionally in the rear half of the bomb bay

Armament (disposable): this can be made up of some 3,629kg (8,000lb) of stores carried in the bomb bay and on two underwing hardpoints, the latter each rated at 454kg (1,000lb); the bomb bay can accommodate six 454kg (1,000lb) bombs on two Avro triple carriers

Electronics and operational equipment: communication and navigation equipment

Powerplant and fuel system: two 3,402kg (7,500lb) thrust Rolls-Royce RA.7 Avon Mk 109 turbojets

Performance: maximum speed 871km/h (541mph) at 12,190m (40,000ft) and 821km/h (510mph) at sea level; initial climb rate 1,097m (3,600ft) per minute; service ceiling 14,630m (48,000ft); range 1,295km (805 miles) with maximum bombload

Weights: empty 12,678kg (27,950lb); normal take-off 19,505kg (43,000lb); maximum take-off 24,925kg (54,950lb)

Dimensions: span 19.49m (63ft 11.5in) without tiptanks; length 19.96m (65ft 6in); height 4.75m (15ft 7in); wing area 89.19m^2 (960sq ft)

British Aerospace Harrier GR.Mk 3

Country of Origin: UK
Type: V/STOL close-support and reconnaissance aircraft
Accommodation: pilot seated on a Martin-Baker Mk 9D ejector seat
Armament (fixed): (optional) two Aden 30mm cannon with 100 rounds per gun carried in place of the underfuselage strakes
Armament (disposable): up to a maximum cleared weight of 2,268kg (5,000lb) though trials have been conducted with up to 3,269kg (8,000lb) of weapons
Electronics and operational equipment: communication and navigation equipment, plus Ferranti FE 541 inertial navigation and attack system, Smiths head-up display, Smiths air-data computer, Marconi ARI.18223 radar-warning receiver, and Ferranti Type 106 laser-ranger and marked-target seeker

Powerplant and fuel system: one 9,752kg (21,500lb) thrust Rolls-Royce (Bristol) Pegasus Mk 103 vectored-thrust turbofan

Performance: maximum speed over 1,186km/h (737mph) or Mach .97 at low altitude; climb to 12,190m (40,000ft) in 2 minutes 22 seconds after VTO; service ceiling more than 15,240m (50,000ft); range 667km (414-mile) radius with 1,361kg (3,000lb) payload after VTO or 5,560km (3,455 miles) for ferrying with one inflight-refuelling

Weights: empty 5,579kg (12,300lb) with pilot; maximum take-off more than 11,340kg (25,000lb)

Dimensions: span 7.70m (25ft 3in), or 9.04m (29ft 8in) with low-drag ferry tips; length 14.27m (46ft 10in); height 3.45m (11ft 4in); wing area 18.68m² (201.1sq ft), or 20.07m² (216sq ft) with ferry tips

British Aerospace Hawk T.Mk 1

Country of Origin: UK
Type: basic and advanced trainer with secondary air-defence and ground-attack roles
Accommodation: pupil and instructor seated in tandem on Martin-Baker Mk 10B rocket-assisted ejector seats
Armament (fixed): optional Aden 30mm cannon in a pod with its ammunition under the fuselage

Armament (disposable): up to a maximum ordnance load of 2,567kg (5,660lb)

Electronics and operational equipment: communication and navigation equipment, plus Ferranti F.195 weapon sight and provision for a centreline reconnaissance pod

Powerplant and fuel system: one 2,359kg (5,200lb) thrust Rolls-Royce/Turboméca Adour Mk 151

Performance: maximum speed 1,038km/h (645mph) or Mach .88 at 3,355m (11,000ft); initial climb rate 2,835m (9,300ft) per minute; service ceiling 15,240m (50,000ft); range 556km (345 miles)

Weights: empty 3,647kg (8,040lb); normal take-off 5,572kg (12,284lb) with weapons as a trainer; maximum take-off 7,750kg (17,085lb)

Dimensions: span 9.39m (30ft 9.75in); length 11.17m (36ft 7.75in) excluding probe; height 3.99m (13ft 1.25in); wing area 16.69m² (179.6sq ft)

British Aerospace Hunter FGA.Mk 9

Country of Origin: UK
Type: fighter and weapons-training aircraft
Accommodation: pilot

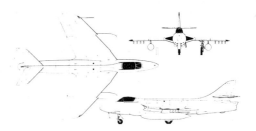

Armament (fixed): four Aden 30mm cannon with
135 rounds per gun in a detachable ventral pack
Armament (disposable): typical underwing stores
are 454kg (1,000lb) bombs plus 227kg (500lb)
bombs

Electronics and operational equipment: communication and navigation equipment, plus ranging radar
Powerplant and fuel system: one 4,604kg (10,150lb) thrust Rolls-Royce RA.28 Avon Mk 207
Performance: maximum speed 1,144km/h (710mph) or Mach .93 at sea level, and 978km/h (620mph) or Mach .94 at high altitude; initial climb rate 2,440m (8,000ft) per minute; service ceiling 15,700m (51,500ft); range 2,965km (1,840 miles) with two 1,046-litre (230-Imp gal) drop-tanks
Weights: empty 5,901kg (13,010lb); normal take-off 8,165kg (18,000lb); maximum take-off 10,886kg (24,000lb) with two 456-litre (100-Imp gal) and two 1,046-litre (230-Imp gal) drop-tanks
Dimensions: span 10.25m (33ft 8in); length 13.98m (45ft 10.5in); height 4.02m (13ft 2in); wing area 32.42m² (349sq ft)

British Aerospace Lightning F.Mk 6

Country of Origin: UK
Type: interceptor fighter
Accommodation: pilot seated on a Martin-Baker ejector seat

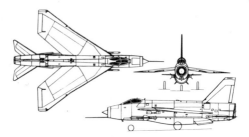

Armament (fixed): two Aden 30mm cannon with 120 rounds per gun in the front portion of the ventral fuel tank
Armament (disposable): up to 144 rockets or six 1,000lb (454kg) HE bombs

Electronics and operational equipment: communication and navigation equipment, plus Ferranti Airpass AI-23S interception radar

Powerplant and fuel system: two 7,420kg (16,360lb) afterburning thrust Rolls-Royce Avon Mk 301 turbojets

Performance: maximum speed 2,112km/h (1,320mph) or Mach 2 at 10,970m (36,000ft); climb to 12,190m (40,000ft) in 2 minutes 30 seconds; service ceiling 16,765m (55,000ft); range 1,287km (800 miles) on internal fuel

Weights: empty 12,717kg (28,041lb); maximum take-off 19,047kg (42,000lb)

Dimensions: span 10.62m (34ft 10in); length 16.84m (55ft 3in) including probe; height 5.97m (19ft 7in); wing area 42.97m² (458.5sq ft)

British Aerospace Nimrod AEW.Mk 3

Country of Origin: UK
Type: airborne early warning aircraft

Accommodation: crew of four on the flightdeck, and normally a tactical crew of six in the cabin
Armament (fixed): none
Armament (disposable): none
Electronics and operational equipment: communication and navigation equipment, plus Marconi

**British
Aerospace
Nimrod MR.2**

multi-mode pulse-Doppler radar in bulged radome at the nose and tail of the aircraft and able to detect air and ship targets at up to 483km (300 miles) despite electronic countermeasures, Loral early warning support measures (EWSM) in two wingtip pods, Cossor Jubilee Guardsman IFF using the same scanners as the radar, passive radio/radar receivers, electronic support measures (ESM) in a pod on top of the heightened vertical tail surfaces, data-link equipment, secure voice communications

Powerplant and fuel system: four 5,507kg (12,140lb) thrust Rolls-Royce RB.168-20 Spey Mk 250 turbofans

Performance: maximum speed 926km/h (575mph); service ceiling 12,800m (42,000ft); range 9,265km (5,755 miles)

Weights: maximum take-off 80,333kg (177,100lb)

Dimensions: span 35.08m (115ft 1in) over EWSM pods; length 41.97m (137ft 8.5in); height 11.58m (38ft); wing area 197m² (2,121sq ft)

British Aerospace Shackleton MR.Mk 3

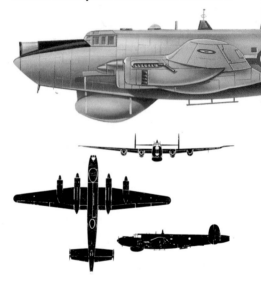

Country of Origin: UK
Type: maritime reconnaissance aircraft
Accommodation: crew of three or four on the flightdeck, gunner and bomb-aimer in the nose and up to five mission crew in the fuselage

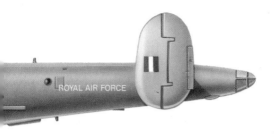

Armament (fixed): two Hispano 20mm cannon on a flexible mount in the nose

Armament (disposable): this is carried in a fuselage weapons bay, and can consist of at least 4,536kg (10,000lb) of torpedoes, depth charges and bombs

Electronics and operational equipment: communication and navigation equipment, plus search radar in a retractable radome in the belly of the aircraft aft of the weapons bay

Powerplant and fuel system: four 1,831kW (2,455hp) Rolls-Royce Griffon 57A inline engines

Performance: maximum speed 486km/h (302mph); cruising speed 407km/h (253mph); initial climb rate 260m (850ft) per minute; service ceiling 5,850m (19,200ft); range typically 5,890km (3,660 miles)

Weights: empty 26,218kg (57,800lb); maximum take-off 44,453kg (98,000lb)

Dimensions: span 36.53m (119ft 10in); length 28.19m (92ft 6in); height 7.11m (23ft 4in); wing area 132m² (1,421sq ft)

British Aerospace Strikemaster Mk 88

Country of Origin: UK
Type: light strike aircraft

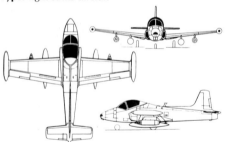

British Aerospace 167 Strikemaster
Accommodation: pilot and co-pilot/pupil
Armament (fixed): two 7.62mm (.3in) FN machine-guns with 550 rounds per gun

Armament (disposable): up to a maximum weight of 1,361kg (3,000lb) with reduced fuel

Electronics and operational equipment: communication and navigation equipment, plus SFOM optical sight, or GM2L reflector sight or Ferranti LFS type 5 gyro sight

Powerplant and fuel system: one 1,424kg (3,140lb) thrust Rolls-Royce (Bristol) Viper Mk 535 turbojet

Performance: maximum speed 772km/h (480mph) at 5,585m (18,000ft) and 724km/h (450mph) at sea level; service ceiling 12,190m (40,000ft); range 233km (145 miles)

Weights: empty 2,810kg (6,195lb); normal take-off 4,808kg (10,600lb) for armament training; maximum take-off 5,215kg (11,500lb)

Dimensions: span 11.23m (36ft 10in) over tiptanks; length 10.27m (33ft 8.5in); height 3.34m (10ft 11.5in); wing area 19.85m^2 (213.7sq ft)

British Aerospace Victor K.Mk 2

Country of Origin: UK
Type: inflight-refuelling tanker
Accommodation: crew of four or five
Armament (fixed): none
Armament (disposable): none

Electronics and operational equipment: communication and navigation equipment, plus radar and inflight-refuelling gear comprising two Flight Refuelling Ltd Mk 20B pods under the wings and one Flight Refuelling Ltd Mk 17B hose-and-drogue unit under the rear fuselage

Powerplant and fuel system: four 9,344kg (20,600lb) thrust Rolls-Royce Conway Mk 201 turbofans, and a total fuel capacity (standard and transfer) of 72,168 litres (15,875 Imp gal) in 10 wing, seven fuselage, two bomb-bay and two non-jettisonable underwing tanks

Performance: maximum speed over 966km/h (600mph) at 12,190m (40,000ft); service ceiling over 15,240m (50,000ft); range 7,403km (4,600 miles)

Weights: empty about 49,896kg (110,000lb); maximum take-off 101,153kg (223,000lb)

Dimensions: span 35.66m (117ft); length 35.03m (114ft 11in); height 8.57m (28ft 1.5in); wing area 204.38m^2 (2,200sq ft)

Canadair CL-41G Tebuan

Country of Origin: Canada
Type: light attack aircraft
Accommodation: pilot and co-pilot seated side-by-side on ejector seats
Armament (fixed): none

Armament (disposable): up to a maximum of 1,814kg (4,000lb) of stores

Electronics and operational equipment: communication and navigation equipment

Powerplant and fuel system: one 1,338kg (2,950lb) thrust General Electric J85-J4 turbojet, and a total internal fuel capacity of 1,135 litres (250 Imp gal) in a five-cell fuselage tank, plus provision for two 182-litre (40-Imp gal) drop-tanks.

Performance: maximum speed 755km/h (470mph) at 8,685m (28,500ft); initial climb rate 1,295m (4,250ft) per minute; service ceiling 12,865m (42,200ft); range 2,220km (1,380 miles) with external fuel

Weights: empty 2,402kg (5,296lb); normal take-off 4,536kg (10,000lb); maximum take-off 5,120kg (11,288lb)

Dimensions: span 11.13m (36ft 6in); length 9.75m (32ft); height 2.81m (9ft 3in); wing area 20.44m² (220sq ft)

CASA C-101EB Aviojet

Country of Origin: Spain
Type: basic and advanced trainer

Accommodation: pupil and instructor seated in tandem on Martin-Baker E10C ejector seats
Armament (fixed): this is accommodated in a quick-change pack under the rear cockpit, and comprises either a DEFA 30mm cannon or two 12.7mm (.5in) FN-Browning machine-guns

Armament (disposable): up to a maximum weight of 1,500kg (3,307lb)

Electronics and operational equipment: communication and navigation equipment, plus alternative pods (photographic reconnaissance, or ECM or laser designation) in place of the gun package in the fuselage

Powerplant and fuel system: one 1,588kg (3,500lb) thrust Garrett TFE 331-2-2J turbofan

Performance: maximum speed 690km/h (429mph) at sea level, and 795km/h (494mph) or Mach 0.71 at 7,620m (25,000ft); service ceiling 12,200m (40,025ft); range 380km (236 miles)

Weights: empty 3,350kg (7,385lb); normal take-off 4,850kg (10,692lb); maximum take-off 5,600kg (12,345lb)

Dimensions: span 10.60m (34ft 9.3in); length 12.25m (40ft 2.25in); height 4.25m (13ft 11.25in); wing area 20.00km² (215.3sq ft)

Cessna Model 318E (A-37B Dragonfly)

Country of Origin: USA
Type: light attack aircraft

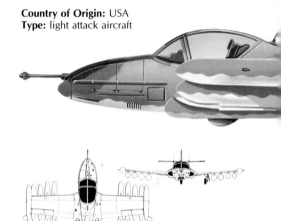

Accommodation: pilot and co-pilot side-by-side
Armament (fixed): one General Electric GAU-2B/A
Minigun in the forward fuselage
Armament (disposable): up to a maximum weight of
2,576kg (5,680lb)

Electronics and operational equipment: communication and navigation equipment, plus Chicago Aerial Industries CA-503 sight

Powerplant and fuel system: two 1,293kg (2,850lb) General Electric J85-GE-17A turbojets

Performance: maximum speed 816km/h (507mph) at 4,875m (16,000ft); cruising speed 787km/h (489mph) at 7,620m (25,000ft); initial climb rate 2,130m (6,990ft) per minute; service ceiling 12,730m (41,765ft); range 740km (460 miles) with 1,860kg (4,100lb) of ordnance, or 1,628km (1,012 miles) with maximum fuel

Weights: empty 2,817kg (6,211lb); maximum take-off 6,350kg (14,000lb)

Dimensions: span 10.93m (35ft 10.5in) over tip-tanks; length 8.62m (28ft 3.25in) excluding probe; height 2.70m (8ft 10.5in); wing area 17.09m² (183.9sq ft)

Dassault-Breguet Alizé

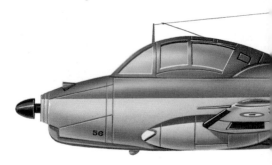

Country of Origin: France
Type: anti-submarine aircraft
Accommodation: pilot seated in the cockpit, and two sensor-system operators in the fuselage
Armament (fixed): none
Armament (disposable): the weapons bay can accept one 500kg (1,102lb) homing torpedo or three 160kg (353lb) depth charges
Electronics and operational equipment: communication and navigation equipment, plus Thomson-CSF DRAA-2B (being replaced by Thomson-CSF Iguane in French aircraft) search radar with its antenna in a retractable 'dustbin' radome in the lower fuselage, and sonobuoys carried in the forward portion of the main landing gear fairings; ECM

equipment is being fitted to the 28 French Alizés retrofitted with Iguane radar

Powerplant and fuel system: one 1,473kW (1,975shp) Rolls-Royce R.DA.7 Dart Mk 21 turboprop.

Performance: maximum speed 520km/h (323mph) at 3,000m (9,845ft); cruising speed 370km/h (230mph); initial climb rate 420m (1,380ft) per minute; service ceiling over 6,250m (20,505ft); range 2,500km (1,553 miles) with standard fuel, and 2,870km (1,783 miles) for ferrying with auxiliary fuel

Weights: empty 5,700kg (12,566lb); maximum take-off 8,200kg (18,078lb)

Dimensions: span 15.6m (51ft 2in); length 13.86m (45ft 6in); height 5m (16ft 4.75in); wing area 36m² (387.51sq ft)

Dassault-Breguet Atlantic Génération 2 (ATL2)

Country of Origin: France
Type: long-range maritime patrol aircraft

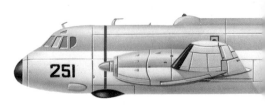

Accommodation: crew of 12, comprising an observer in the nose, flight crew of three in the cockpit, and mission crew of six in the tactical compartment, and two observers
Armament (fixed): none
Armament (disposable): up to a maximum weight of 3,000kg (6,600lb)
Electronics and operational equipment: communication and navigation equipment, plus Thomson CSF Iguane search radar in a retractable radome forward of the stores bay, SAT/TRT forward-looking infra-red (FLIR) sensor
Powerplant and fuel system: two 4,225kW (5,665shp) Rolls-Royce Tyne RTy.20 Mk 21

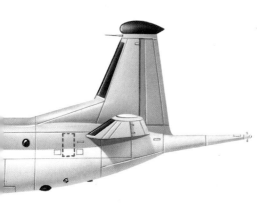

Performance: maximum speed 645km/h (400mph) at optimum altitude, and 592km/h (368mph) at sea-level; cruising speed 315km/h (195mph) between sea level and 1,525m (5,000ft); initial climb rate 610m (2,000ft) per minute; service ceiling 9,145m (30,000ft); range 9,075km (5,640 miles), or a patrol endurance of 5 hours at a radius of 1,850km (1,150 miles)

Weights: empty 25,300kg (55,776lb); normal take-off 43,900kg (96,781lb); maximum take-off 46,200kg (101,852lb)

Dimensions: span 37.42m (122ft 9.25in) with tip pods; length 33.63m (110ft 4in); height 10.89m (35ft 8.75in); wing area 120.34m² (1,295sq ft)

Dassault-Breguet/Dornier Alpha Jet

Country of Origin: France/West Germany
Type: advanced jet trainer and battlefield close-support/reconnaissance aircraft

Accommodation: pupil and instructor seated in tandem on Martin-Baker AJRM4 (French aircraft)
Armament (fixed): one Mauser 27mm or DEFA 30mm cannon

Armament (disposable): up to a maximum of 2,500kg (5,511lb)

Electronics and operational equipment: communication and navigation equipment

Powerplant and fuel system: two 1,350kg (2,976lb) thrust SNECMA/Turboméca Larzac 04-C5 turbofans

Performance: maximum speed 1,005km/h (624mph) or Mach .85 at 3,050m (10,000ft); service ceiling 14,630m (48,000ft); range 425km (264 miles)

Weights: normal take-off 5,000kg (11,023lb) as a trainer; maximum take-off 7,500kg (16,534lb) for attack

Dimensions: span 9.11m (29ft 10.75in); length 12.29m (40ft 3.75in) as a trainer, and 13.23m (43ft 5in) as an attack aircraft; wing area 17.5m² (188.4sq ft)

Dassault-Breguet Etendard IVM

Country of Origin: France
Type: shipboard strike aircraft
Accommodation: pilot seated on a Hispano-built Martin-Baker SEMMB CM4A lightweight ejector seat
Armament (fixed): two DEFA 30mm cannon with 125 rounds per gun in the lower lips of the engine inlets
Armament (disposable): this is carried on four underwing hardpoints up to a weight of 1,360kg (3,000lb); stores which can be carried are the AIM-9 Sidewinder air-to-air missile, the AS 30 air to surface missile various types of rocket-launcher pod, 225kg (496lb) bombs and 410kg (904lb) bombs
Electronics and operational equipment: communication and navigation equipment, plus Electronique Marcel Dassault Aida ranging radar
Powerplant and fuel system: one 4,400kg (9,700lb) thrust SNECMA Atar 08B turbojet and a total inter-

nal fuel capacity of 3,300 litres (726 Imp gal) in wing and fuselage tanks, plus provision for two 600-litre (132-Imp gal) drop tanks; inflight-refuelling capability

Performance: maximum speed 1,100km/h (684mph) or Mach .9 at sea level, and 1,085km/h (674mph) or Mach 1.02 at 11,000m (36,090ft); initial climb rate 6,000m (19,685ft) per minute; service ceiling 15,000m (49,215ft); range 1,600km (994 miles) with external weapons and 3,000km (1,864 miles) for ferrying with underwing tanks

Weights: empty 6,125kg (13,503lb); normal take-off 8,170kg (18,011lb); maximum take-off 10,275kg (22,652lb)

Dimensions: span 9.60m (31ft 5.75in); length 14.40m (47ft 3in); height 4.26m (14ft 0in); wing area 29.0m² (312.16sq ft)

Dassault-Breguet M.D.450 Ouragan

Country of Origin: France
Type: fighter-bomber
Accommodation: pilot seated on a Martin-Baker ejector seat
Armament (fixed): four Hispano 404 Modèle 50 20mm cannon with 125 rounds per gun in the underside of the nose
Armament (disposable): this is carried on underwing hardpoints, and can comprise two 500kg (1,102lb) bombs, or 16 105mm (4.13in) rockets, or eight 105mm (4.13in) rockets and two 455-litre (100-Imp gal) napalm tanks
Electronics and operational equipment: communication and navigation equipment

Powerplant and fuel system: one 2,268kg (5,000lb) thrust Hispano-built Rolls-Royce Nene Mk 104B turbojet

Performance: maximum speed 940km/h (584mph) or Mach .72 at sea level, and 830km/h (516mph) or Mach .78 at 12,000m (39,370ft); initial climb rate 2,400m (7,875ft) per minute; service ceiling 13,000m (42,650ft); range 920km (572 miles) without external stores

Weights: empty 4,140kg (9,127lb); maximum take-off 7,900kg (17,416lb)

Dimensions: span 13.16m (43ft 2in) over tiptanks, length 10.74m (25ft 2.75in); height 4.14m (13ft 7in); wing area 23.80m^2 (256.18sq ft)

Dassault-Breguet M.D.452 Mystère IVA

Country of Origin: France
Type: fighter-bomber
Accommodation: pilot seated on a Martin-Baker ejector seat
Armament (fixed): two DEFA 30mm cannon in the underside of the nose, and provision for a Matra 101bis retractable underfuselage pack of 55 Brandt air-to-air rockets
Armament (disposable): this is carried on four underwing hardpoints up to a weight of 900kg (1,984lb); weapons which can be carried include bombs, or 12 T-10 air-to-surface rockets, or two Matra launchers each with 19 68mm (2.68in) rockets

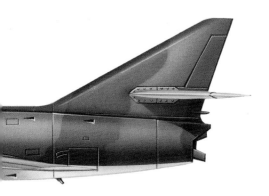

Electronics and operational equipment: communication and navigation equipment

Powerplant and fuel system: one 3,500kg (7,716lb) thrust Hispano-Suiza Verdon 350 turbojet

Performance: maximum speed 1,120km/h (696mph) or Mach .91 at sea level, and 990km/h (615mph) or Mach .93 at 12,000m (39,370ft); initial climb rate 2,700m (8,860ft) per minute; service ceiling 15,000m (49,215ft); range 915km (569 miles) without external stores

Weights: empty 5,870kg (12,941lb); maximum take-off 9,500kg (20,944lb)

Dimensions: span 11.12m (36ft 5.75in); length 12.85m (42ft 2in); height 4.60m (15ft 1in); wing area 32m² (344.46sq ft)

Dassault-Breguet Mirage IIIE

Country of Origin: France
Type: fighter-bomber

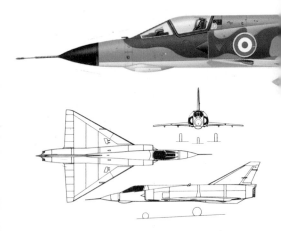

Accommodation: pilot seated on a Hispano-built Martin-Baker RM4 ejector seat
Armament (fixed): two DEFA 552A 30mm cannon with 125 rounds per gun
Armament (disposable): up to a maximum of about 2,270kg (5,000lb)

Electronics and operational equipment: communication and navigation equipment, plus CSF Cyrano II fire-control radar (air-to-air and air-to-surface), CSF 97 sighting system, Marconi Doppler radar and other specialized avionics

Powerplant and fuel system: one 6,200kg (13,670lb) afterburning thrust SNECMA Atar 9C

Performance: maximum speed clean 2,350km/h (1,460mph) or Mach 2.2 at 12,000m (39,370ft) service ceiling 17,000m (55,775ft) without rocket motor; range 1,200km (745 miles)

Weights: empty 7,050kg (15,540lb); normal take-off 9,600kg (21,165lb); maximum take-off 13,700kg (30,200lb)

Dimensions: span 8.22m (26ft 11.5in); length 15.03m (49ft 3.5in); height 4.50m (14ft 9in); wing area 35m² (376.75sq ft)

Dassault-Breguet Mirage 5

Country of Origin: France
Type: ground-attack aircraft and interceptor
Accommodation: pilot seated on a Hispano-built Martin-Baker RM4 ejector seat
Armament (fixed): two DEFA 552A 30mm cannon with 125 rounds per gun
Armament (disposable): this is carried on one under-fuselage hardpoint, rated at 1,180kg (2,600lb), and six underwing hardpoints, the inner tandem pair rated at 1,680kg (3,704lb) each and the outer pair at 168kg (370lb) each, to a maximum of more than 4,000kg (8,818lb) with the use of multiple launchers
Electronics and operational equipment: communication and navigation equipment, including an inertial navigation system and navigation system, plus either an Aida II radar and CSF LT 102 or TAV 34 laser-ranging equipment, or an Agave multi-mode radar

Powerplant and fuel system: one 6,200kg (13,670lb) afterburning thrust SNECMA Atar 9C turbojet

Performance: maximum speed clean 2,350km/h (1,460mph) or Mach 2.2 at 12,000m (39,370ft), and 1,390km/h (863mph) or Mach 1.13 at sea level; cruising speed 955km/h (593mph) or Mach .9 at 11,000m (36,090ft); climb to 11,000m (36,090 ft) in 3 minutes; service ceiling 17,000m (55,775ft); range 1,300km (808-mile) hi-lo-hi radius or 650km (404-mile) lo-lo-lo radius with 907kg (2,000lb) bombload, and 4,000km (2,485 miles) for ferrying

Weights: empty 6,600kg (14,550lb); normal take-off 9,600kg (21,165lb); maximum take-off 13,700kg (30,200lb)

Dimensions: span 8.22m (26ft 11.5in); length 15.55m (51ft .25in); height 4.50m (14ft 9in); wing area 35m² (376.75sq ft)

Dassault-Breguet Mirage 50

Country of Origin: France
Type: multi-mission fighter
Accommodation: pilot seated on a Hispano-built Martin-Baker RM4 ejector seat
Armament (fixed): two DEFA 552A 30mm cannon with 125 rounds per gun
Armament (disposable): this is carried on one under-fuselage hardpoint, rated at 1,180kg (2,600lb), and six underwing hardpoints, the inner tandem pair rated at 1,680kg each and the outer pair at 168kg (370lb) each, to a maximum of more than 4,000kg (8,818lb) with the use of multiple launchers
Electronics and operational equipment: communication and navigation equipment, plus CSF-ESD Agave radar (Magic missile installation) or Thomson-CSF Cyrano IVM multi-mode radar (R.530 missile installation), Sagem inertial navigation system and Thomson-CSF head-up display

Powerplant and fuel system: one 7,200kg (15,873lb) afterburning thrust SNECMA Atar 9K-50 turbojet, and a total internal fuel capacity of 3,475 litres (764 Imp gal) plus 1,225 litres (269 Imp gal) in drop-tanks

Performance: maximum speed clean 2,350km/h (1,460mph) or Mach 2.2 at 12,000m (39,370ft), and 1,390km/h (863mph) or Mach 1.13 at sea level; initial climb rate 11,100m (36,415ft) per minute; service ceiling 18,000m (59,055ft); range 630km (391-mile) combat radius at low level with two 400kg (882lb) bombs

Weights: empty 7,150kg (15,765lb); normal take-off 9,900kg (21,825lb); maximum take-off 13,700kg (30,200lb)

Dimensions: span 8.22m (27ft); length 15.56m (51ft .5in); height 4.5m (14ft 9in)

Dassault-Breguet Mirage 2000

Country of Origin: France
Type: interceptor and air-superiority fighter
Accommodation: pilot seated on a Martin-Baker
F10Q ejector seat
Armament (fixed): two DEFA 554 30mm cannon
with 125 rounds per gun

Armament (disposable): this is carried on nine hard-points, one under the fuselage rated at 1,800kg (3,968lb), four under the wing roots each rated at 400kg (882lb), and four under the wings to a maximum weight of more than 6,000kg (13,228lb)

Powerplant and fuel system: one 9,000kg (19,840lb) SNECMA M53-5 bleed-turbojet

Performance: maximum speed more than 2,350km/h (1,460mph) or Mach 2.2 at 12,000m (39,370ft), and 1,110km/h (690mph) or Mach .9 at sea level with bombs; initial climb rate more than 18,000m (59,055ft) per minute; service ceiling 20,000m (65,615ft); range more than 1,800km (1,118 miles) with two 1,700-litre (374-Imp gal) drop-tanks

Weights: empty 7,400kg (16,315lb); maximum take-off 16,500kg (36,375lb)

Dimensions: span 9m (29ft 6in); length 14.35m (47ft 1in); wing area 41m² (441.3sq ft)

Dassault-Breguet Mirage F.1C

Country of Origin: France
Type: multi-role fighter and attack aircraft
Accommodation: pilot seated on a SEM Martin-Baker F1RM4 ejector seat
Armament (fixed): two DEFA 553 30mm cannon with 135 rounds per gun
Armament (disposable): this is carried on one under-fuselage, four underwing and two wingtip hardpoints (the last suitable only for Matra 550 Magic or AIM-9 Sidewinder air-to-air missiles), up to maximum weight of 4,000kg (8,818lb)
Electronics and operational equipment: communication and navigation equipment, plus Thomson-CSF Cyrano IV fire-control radar, Sagem Uliss 47 inertial navigation system, CSF head-up display, Doppler radar, laser-rangefinder and terrain-avoidance radar

Powerplant and fuel system: one 7,200kg (15,873lb) afterburning thrust SNECMA Atar 9K-50 turbojet

Performance: maximum speed 2,350km/h (1,460mph) or Mach 2.2 at 12,000m (39,370ft) and 1,470km/h (913mph) or Mach 1.2 at sea-level; initial climb rate 12,780m (41,930ft) per minute; service ceiling 20,000m (65,615ft); range 600km (373-mile) lo-lo-lo radius with six 250kg (551lb) bombs, or 425km (264-mile) hi-lo-hi radius with 14 250kg (551lb) bombs

Weights: empty 7,400kg (16,314lb); normal take-off 10,900kg (24,030lb); maximum take-off 16,200kg (35,714lb)

Dimensions: span 8.40m (27ft 6.75in); length 15m (49ft 2.5in); height 4.5m (14ft 9in); wing area 25m² (269.1sq ft)

Dassault-Breguet Super Etendard

Country of Origin: France
Type: carrier-based strike fighter
Accommodation: pilot seated on a Hispano-built Martin-Baker SEMMB CM4A ejector seat
Armament (fixed): two DEFA 30mm cannon, plus up to 2,100kg (4,630lb) bombload
Electronics and operational equipment: communication and navigation equipment, plus Thomson-

CSF/ESD Agave lightweight multi-function radar, Sagem-Kearfott ETNA inertial platform, Thomson-CSV VE-120 head-up display, Crouzet Type 66 air-data computer and Crouzet Type 97 navigation display and armament control system

Powerplant and fuel system: one 5,000kg (11,023lb) thrust SNECMA Atar 8K-50 turbojet

Performance: maximum speed about 1,065km/h (662mph) or Mach 1.0 at 11,000m (36,090ft), and 1,180km/h (733mph) or Mach .96 at sea level; initial climb rate 6,000m (19,685ft) per minute; service ceiling 13,700m (44,950ft); range 850km (528-mile) radius with AM.39 Exocet missile

Weights: empty 6,500kg (14,330lb); normal take-off 9,450kg (20,835lb); maximum take-off 12,000kg (26,455lb)

Dimensions: span 9.6m (31ft 6in); length 14.31m (46ft 11.5in); height 3.86m (12ft 8in); wing area 24.8m² (267sq ft)

Dassault-Breguet Super Mirage 4000

Country of Origin: France
Type: multi-role combat aircraft
Accommodation: pilot seated on a Martin-Baker F10R ejector seat
Armament (fixed): two 30mm cannon (probably DEFA 554 weapons)
Armament (disposable): this is carried on 11 hardpoints (one under the fuselage, four under the wing roots, and six under the wings) to a maximum of more than 8,000kg (17,637lb)
Electronics and operational equipment: communication and navigation equipment, plus Thomson-CSF/ESD RDI pulse-Doppler radar, Sagem Uliss 52 inertial platform, Crouzet Type 80 digital air-data computer, Thomson-CSF VE-130 head-up display, and an automated digital stores management/weapon delivery system

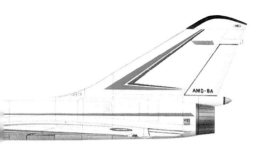

Powerplant and fuel system: two 9,700kg (21,385lb) afterburning thrust SNECMA M53-P2 bleed-turbojet, and a total internal fuel capacity of about 11,400 litres (2,508 Imp gal) in fin, fuselage and wing tanks, plus provision for three 2,500-litre (550-Imp gal) drop-tanks (one under the fuselage and two under the wings)

Performance: maximum speed more than 2,450km/h (1,522mph) or Mach 2.3 at 12,000m (39,370ft); initial climb rate 18,300m (60,040ft) per minute; service ceiling 20,000m (65,615ft); range more than 2,000km (1,243 miles) with external fuel

Weights: normal take-off 16,100kg (35,495lb)

Dimensions: span 12m (39ft 4.5in); length 18.7m (61ft 4.5in); wing area 73m² (785.8sq ft)

Dassault-Breguet Super Mystère B2

Country of Origin: France
Type: fighter-bomber and trainer
Accommodation: pilot seated on a Martin-Baker ejector seat
Armament (fixed): two DEFA 552 30mm cannon in the underside of the nose, and provision for a retractable pack of 35 68mm (2.68in) rockets
Armament (disposable): this is carried on two underwing hardpoints, up to a weight of 1,000kg (2,205lb); typical weapons are 500kg (1,102lb) bombs, Matra launchers for 18 68mm (2.68in) rockets, 12 5in (127mm) air-to-surface rockets, napalm tanks and Matra air-to-air missiles

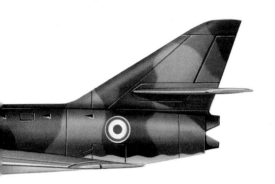

Electronics and operational equipment: communication and navigation equipment, plus ranging radar
Powerplant and fuel system: one 4,500kg (9,921lb) afterburning thrust SNECMA Atar 101G turbojet
Performance: maximum speed 1,040km/h (646mph) or Mach .85 at sea level, and 1,195km/h (743mph) or Mach 1.145 at 11,000m (36,090ft); initial climb rate 5,335m (17,505ft) per minute; service ceiling 17,000m (55,775ft); range 870km (540 miles) without external stores
Weights: empty 6,930kg (15,278lb); maximum take-off 10,000kg (22,046lb)
Dimensions: span 10.52m (34ft 6in); length 14.13m (46ft 4.25in); height 4.55m (14ft 11in); wing area 35in² (376.75sq ft)

Douglas EA-3B Skywarrior

Country of Origin: USA
Type: electronic reconnaissance aircraft
Accommodation: crew of three on the flightdeck, and a mission crew of four in the cabin replacing the bomb-bay

Armament (fixed): two 20mm cannon in a radar-controlled rear barbette
Armament (disposable): none
Electronics and operational equipment: communication and navigation equipment, plus ECM gear, side-looking airborne radar (SLAR), forward-looking radar, chaff dispensers and infra-red sensors

Powerplant and fuel system: two 4,763kg (10,500lb) thrust Pratt & Whitney J57-P-10 turbojets, and a total internal fuel capacity of 19,025 litres (4,185 Imp gal) in integral wing and two fuselage tanks; inflight-refuelling capability

Performance: maximum speed 982km/h (610mph) at 3,050m (10,000ft); cruising speed 837km/h (520mph); service ceiling 12,495m (41,000ft); range 4,665km (2,900 miles)

Weights: empty 17,876kg (39,409lb); normal take-off 31,751kg (70,000lb); maximum take-off 37,195kg (82,000lb)

Dimensions: span 22.1m (72ft 6in); length 23.27m (76ft 4in); height 6.95m (22ft 9.5in); wing area 75.44m² (812sq ft)

EMBRAER AT-26 Xavante

Country of Origin: Brazil
Type: advanced trainer and light attack aircraft
Accommodation: pupil and instructor in tandem on Martin-Baker Mk O4A lightweight ejector seats
Armament (fixed): none
Armament (disposable): up to 2,496kg (5,503lb) can be carried on six underwing hardpoints, the inner four rated at 454kg (1,000lb) each and the outer two at 340kg (750lb) each; all offensive stores are of Brazilian manufacture, and consist of 45, 227 and 250kg (100, 500 and 551lb) bombs, gun pods with twin 7.62mm (.3in) machine-guns, photographic reconnaissance pods, pods for seven or 19 70mm (2.75in) rockets or 36 37mm (1.46in) rockets, and drop tanks
Electronics and operational equipment: communication and navigation equipment

Powerplant and fuel system: one 1,134kg (2,500lb) Rolls-Royce Viper 20 Mk 540 turbojet

Performance: maximum speed 867km/h (539 mph) without stores; crusing speed 797km/h (495mph) without stores; initial climb rate 1,844m (6,050ft) per minute without stores; service ceiling 14,325m (47,000ft) without stores; range 1,850km (1,150 miles) clean as a trainer, or a 130km (90-mile) combat radius with full underwing stores load

Weights: empty 2,685kg (5,90lb) as a trainer, or 2,558kg (5,640lb) for attack; normal take-off 4,577kg (10,090lb) as a trainer, or 4,447kg (9,803lb) for attack; maximum take-off 5,216kg (11,500lb) for attack

Dimensions: span 10.854m (35ft 7.25in) over tip-tanks; length 10.673m (35ft .25in); height 3.72m (12ft 2in); wing area 19.35m² (208.3sq ft)

91

EMBRAER P-95 Bandeirulha

Country of Origin: Brazil
Type: maritime reconnaissance aircraft
Accommodation: crew of two on the flightdeck, plus
a mission crew of three in the cabin

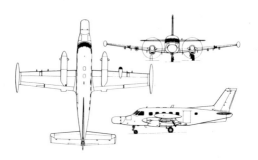

Armament (fixed): none

Armament (disposable): an assortment of stores can be accommodated on four underwing hardpoints

Electronics and operational equipment: Litton LN-33 inertial navigation system; AIL SPAR-1 (AN/APS-128) search radar able to pick up low-profile targets

Powerplant and fuel system: two 559-kW (750-shp) Pratt & Whitney Aircraft of Canada PT6A-34

Performance: maximum speed 404km/h (251mph) at 3,050m (10,000ft); service ceiling 8,230m (27,000ft); range 2,075km (1,290 miles)

Weights: empty 3,403kg (7,502lb); maximum take-off 7,000kg (15,432lb)

Dimensions: span 15.96m (52ft 4.5in) over tiptanks; length 14.83m (48ft 7.9in); height 4.74m (15ft 6.5in); wing area 29m² (312.2sq ft)

Fairchild Republic A-10A Thunderbolt II

Country of Origin: USA
Type: close-support aircraft
Accommodation: pilot seated on a Douglas ACES II ejector seat
Armament (fixed): one General Electric GAU-8/A Avenger rotary-barrel 30mm cannon in the forward fuselage with 1,174 rounds
Armament (disposable): the maximum external load with reduced fuel is 7,258kg (16,000lb), reducing to 5,505kg (14,340lb) with a maximum fuel load; typical loads are 28 227kg (500lb) Mk 82 free-fall or retarded bombs, six 907kg (2,000lb) Mk 84 general-purpose bombs, eight BLU-1 or BLU-27/B incendiary bombs, 20 Rockeye II cluster bombs, 16 CBU-52 or CBU-71 bomb dispensers, six AGM-65 Maverick air-to-surface missiles, Mk 82 and Mk 84 laser-guided bombs, Mk 84 electro-optically guided bombs, two SUU-23/A 20mm cannon pods, and several other stores

Electronics and operational equipment: communication and navigation equipment, plus Kaiser head-up display, weapon-delivery package, AN/AAS-35 'Pave Penny' laser designator pod, AN/ALR-46(V) radar-warning receiver, AN/ALQ-119 ECM pods and other active or passive electronic countermeasures

Powerplant and fuel system: two 4,111kg (9,065lb) thrust General Electric TF34-GE-100 turbofans

Performance: maximum speed 706km/h (439mph) at sea-level in clean condition; cruising speed 623km/h (387mph) at 1,525m (5,000ft); initial climb rate 1,830m (6,000ft) per minute; range 463km (288-mile) close-air support radius with a 1.7-hour loiter, or 3,950km (2,455 miles) for ferrying against a 93km/h (58mph) wind

Weights: empty 11,322kg (24,960lb); maximum take-off 22,680kg (50,000lb)

Dimensions: span 17.53m (57ft 6in); length 16.26m (53ft 4in); height 4.47m (14ft 8in); wing area 47.01m² (506sq ft)

Fairchild Republic F-105D Thunderchief

Country of Origin: USA

Type: tactical fighter-bomber

Accommodation: pilot only, seated on a Martin-Baker ejector seat

Armament (fixed): one General Electric M61A-1 Vulcan 20mm rotary-barrel cannon with 1,029 rounds in the port side of the nose

Armament (disposable): this is carried in an internal bay in the lower fuselage and on five hardpoints (one under the fuselage and four under the wings) up to a maximum weight in excess of 6,350kg (14,000lb); typical load is one 2,461 litre (540 Imp gal) drop tank on the centreline hardpoint with one 1,703 litre (375 Imp gal) drop tank on one inner underwing point and one nuclear weapon on the other

Electronics and operational equipment: communications and navigation equipment plus a General Electric FC-5 integrated automatic flight-control and fire-control system and (in some aircraft) the Thunderstick II all-weather blind-attack bombing system

Powerplant and fuel system: one 12,030kg (26,500lb) afterburning thrust General Electric J75-P-19W turbojet

Performance: maximum speed 2,230km/h (1,385mph) or Mach 2.1 at 10,970m (36,000ft) and 1,375km/h (855mph) or Mach 1.125 at sea level; cruising speed 1,344km/h (835mph) at optimum altitude; initial climb rate 10,515m (34,500ft) per minute; service ceiling 15,850m (52,000ft); combat radius 1,448km (900 miles); ferry range 3,541km (2,200 miles)

Weights: empty 12,474kg (27,500lb); maximum take-off 23,968kg (52,840lb)

Dimensions: span 10.65m (34ft 11.25in); length 19.51m (64ft); height 5.99m (19ft 8in); wing area 35.77m² (385m²)

FMA IA 58A Pucará

Country of Origin: Argentina
Type: two-seat counterinsurgency aircraft
Accommodation: pilot and co-pilot in tandem on Martin-Baker Mk APO6A ejector seats
Armament (fixed): two 20mm Hispano-Suiza HS-804 cannon with 270 rounds each, and four 7.62mm (.3in) FN-Browning machine-guns

A-507

Armament (disposable): a total of 1,620kg (3,571lb) can be carried on one underfuselage and two underwing hardpoints, the former rated at 1,000kg (2,205lb) and the latter each at 500kg (1,102lb)

Electronics and operational equipment: communication and navigation equipment, plus and AN/AWE-1 stores programmer

Powerplant and fuel system: two 761-kW (1,022-ehp) Turboméca Astazou XVIG turboprops

Performance: maximum speed 500km/h (310mph) at 3,000m (9,845ft); cruising speed 480km/h (298mph) at 6,000m (19,685ft); initial climb rate 1,080m (3,545ft) per minute; service ceiling 10,000m (32,810ft); range 3,040km (1,889 miles)

Weights: empty 4,037kg (8,900lb); maximum take-off 6,800kg (14,991lb)

Dimensions: span 14.5m (47ft 6.75in); length 14.25m (46ft 9in); height 5.36m (17ft 7in); wing area 30.3m² (326.1sq ft)

General Dynamics F-16A Fighting Falcon

Country of Origin: USA
Type: lightweight air-combat fighter
Accommodation: pilot seated on a Douglas ACES II ejector seat
Armament: one General Electric M61A1 Vulcan rotary-barrel 20mm cannon in the port wing/fuselage fairing with 500 rounds. Disposable load is 9,276kg (20,450lb).
Electronics and operational equipment: communication and navigation equipment, plus Westinghouse AN/APG-66 pulse-Doppler range and angle track radar (with look-down and look-up ranges of 56km (35 miles) and 74km (46 miles) respectively), Dalmo Victor AN/ALR radar-warning receiver, Sperry central air-data computer, Singer-Kearfott SKN-2400 (modified) inertial navigation system, Marconi head-up display, Kaiser radar electro-optional display, Delco fire-control computer, Martin-Marietta 'Pave Penny' laser-tracking pod, Martin-Marietta LANTIRN FLIR pod, Westinghouse AN/

ALQ-119 and AN/ALQ-131 ECM pods and other electronic countermeasures equipment

Powerplant and fuel system: one 11,340kg (25,000lb) afterburning thrust Pratt & Whitney F100-PW-200 turbofan

Performance: maximum speed more than 2,124km/h (1,320mph) or Mach 2 at 12,190m (40,000ft); service ceiling more than 15,240m (50,000ft); range more than 925km (575-mile) combat radius, or more than 3,890km (2,145 miles) for ferrying with internal and external fuel

Weights: empty 8,065kg (17,780lb); normal take-off 11,633kg (25,647lb); maximum take-off 16,057kg (35,400lb)

Dimensions: span 9.45m (31ft) over missile rails; length 15.09m (49ft 5.9in); height 5.09m (16ft 8.5in); wing area 27.87m² (300sq ft)

General Dynamics F-106A Delta Dart

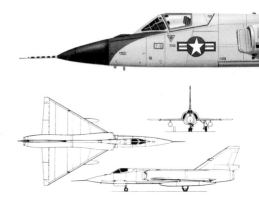

Country of Origin: USA
Type: interceptor fighter
Accommodation: pilot only, seated on an ejector seat
Armament (fixed): one General Electric M61A-1 Vulcan 20mm rotary-barrel cannon
Armament (disposable): this is carried in three internal weapons bays, and comprises one Douglas AIR-2A Genie or AIR-2B Super Genie nuclear-warhead

air-to-air rocket plus two or four Hughes AIM-4F or AIM-4G Super Falcon air-to-air missiles

Electronics and operational equipment: communication and navigation equipment, plus Hughes MA-1 interception and fire-control system tied in by data-link to the SAGE (Semi-Automatic Ground Environment) system within the NORAD (North American Air Defence) network

Powerplant and fuel system: one 11,113kg (24,500lb) afterburning thrust Pratt & Whitney

Performance: maximum speed 2,455km/h (1,525mph) or Mach 2.3 at 10,970m (36,000ft); initial climb rate about 9,145m (30,000ft) per minute; service ceiling 17,325m (57,000ft); range 1,850km (1,150 miles)

Weights: empty 10,726kg (23,646lb); maximum take-off 17,554kg (38,700lb)

Dimensions: span 11.67m (38ft 3.5in); length 21.56m (70ft 8.75in); height 6.18m (20ft 3.25in); wing area 58.65m² (631.3sq ft)

103

General Dynamics F-111F

Country of Origin: USA

Type: variable-geometry multi-role fighter

Accommodation: crew of two seated side-by-side in a McDonnell Douglas escape capsule with a 18,144kg (40,000lb) thrust rocket motor

Armament (fixed): one General Electric M61A1 Vulcan rotary-barrel 20mm cannon (optional) in the weapons bay

Armament (disposable): this is carried in an internal weapons bay and on four underwing hardpoints; the internal weapons bay can carry two (only one if cannon pack is fitted) 340kg (750lb) B43 nuclear free-fall bombs, and the maximum ordnance load is 14,288kg (31,500lb); weapons that can be carried on the underwing hardpoints are the M117 340kg (750lb) retarded bomb, Mk 84 907kg (2,000lb) free-fall or retarded bomb, Mk 82 227kg (500lb) free-fall or retarded bomb, SUU-30 fragmentation-bomb dispenser, CBU-30 and CBU-38 bomb dispensers

Electronics and operational equipment: communication and navigation equipment, plus General Elec-

tric AN/APQ-119 attack and navigation radar, Texas Instruments AN/APQ-128 terrain-following radar, IBM AN/ASQ-133 digital fire-control computer, General Electric AN/ASG-25 optical display sight, electronic countermeasures, and Ford AN/AVQ-26 'Pave Tack' target acquisition and designation pod

Powerplant and fuel system: two 11,385kg (25,100lb) afterburning thrust Pratt & Whitney TF30-PW-100 turbofans, plus provision for two underwing drop-tanks; inflight-refuelling capability

Performance: maximum speed 2,655km/h (1,650mph) or Mach 2.5 at high altitude, and 1,473km/h (915mph) or Mach 1.2 at sea-level; service ceiling 18,290m (60,000ft); range more than 4,707km (2,925 miles) with maximum internal fuel

Weights: empty 21,398kg (47,175lb); maximum take-off 45,359kg (100,000lb)

Dimensions: span 19.m (63ft) spread and 9.74m (31ft 11.4in) swept; length 22.4m (73ft 6in); height 5.22m (17ft 1.4in); wing area 48.77m² (525sq ft) spread and 61.07m² (657.3sq ft) swept

Grumman A-6E/TRAM Intruder

Country of Origin: USA

Type: shipboard attack bomber

Accommodation: pilot and bombardier seated in echelon on Martin-Baker GRU17 ejector seats

Armament (fixed): none

Armament (disposable): this is carried on one under-fuselage and four underwing hardpoints, each rated at 1,633kg (3,600lb) up to a maximum weight of 8,165kg (18,000lb)

Electronics and operational equipment: communication and navigation equipment, plus Norden AN/APQ-148 multi-mode radar (navigation, target identification and tracking, airborne moving target indication/AMTI, and terrain-clearance and terrain-avoidance), IBM AN/ASQ-133 digital navigation/attack computer system, Kaiser AN/AVA-1 multi-mode electronic display, Litton AN/ASN-92 inertial navi-

gation system, CNI (communication, navigation and identification) system, Philco Ford 'Pave Knife' laser designator pod, and an undernose TRAM (Target Recognition and Attack Multi-sensor) package with infra-red and laser sensors

Powerplant and fuel system: two 4,218kg (9,300lb) thrust Pratt & Whitney J52-P-8B turbojets

Performance: maximum speed 1,037km/h (644mph) at sea level; cruising speed 763km/h (474mph) at optimum altitude; initial climb rate 2,323m (7,620ft) per minute; service ceiling 2,925km (42,400ft); range 1,627km (1,011 miles) with maximum payload, or 4,401km (2,735 miles) for ferrying with maximum internal and external fuel

Weights: empty 12,093kg (26,660lb); maximum take-off 27,397kg (60,400lb)

Dimensions: span16.15m (53ft); length 16.69m (54ft 9in); height 4.93m (16ft 2in); wing area 49.1m² (528.9sq ft)

Grumman E-2C Hawkeye

Country of Origin: USA
Type: shipboard early-warning aircraft
Accommodation: crew of two on the flightdeck, and a tactical team of three (combat information centre officer, air-control officer and radar operator) in the cabin
Armament (fixed): none
Armament (disposable): none
Electronics and operational equipment: communication and navigation equipment, plus General Electric AN/APS-125 search radar (with a range against small targets of 185km (115 miles), in a Randtron AN/APA-171 rotodome, Litton AN/ALR-59 passive detection system, Hazeltine AN/APA-172 control indicator group, Litton AN/ASN-92 carrier aircraft

inertial navigation system, Conrac air-data computer, Litton L-304 computer system, AN/APN-153(V) Doppler navigation and other systems

Powerplant and fuel system: two 3,661kW (4,810ehp) Allison T56-A-425 turboprops

Performance: maximum speed 602km/h (374mph); cruising speed 587km/h (365mph); service ceiling 9,390m (30,800ft); range 320km (200-mile) radius with a 4-hour patrol, or 2,583km (1,605 miles) for ferrying

Weights: empty 17,211kg (37,945lb); maximum take-off 23,503kg (51,817lb)

Dimensions: span 24.56m (80ft 7in); length 17.54m (57ft 6.75in); height 5.58m (18ft 3.75in); wing area 65.03m² (700sq ft)

Grumman EA-6B Prowler

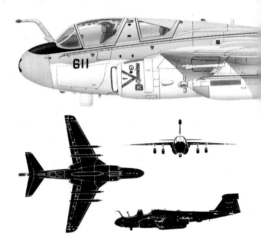

Country of Origin: USA
Type: shipboard electronic countermeasures aircraft
Accommodation: crew of four (pilot, navigator and two electronics officers) seated on Martin-Baker GRUEA 7 ejector seats
Armament (fixed): none
Armament (disposable): none

Electronics and operational equipment: communication and navigation equipment, plus the Raytheon AN/ALQ-99 tactical jamming system

Powerplant and fuel system: two 5,080kg (11,200lb) thrust Pratt & Whitney J52-P-408 turbos

Performance: maximum speed 1,002km/h (623mph) at sea level; cruising speed 774km/h (481mph) at optimum altitude; initial climb rate 3,057m (10,030ft) per minute; service ceiling 11,580m (38,000ft); range 535km (332-mile) radius with a 1-hour loiter, or 3,254km (2,022 miles) for ferrying

Weights: empty 14,588kg (32,162lb); normal take-off 24,703kg (54,461lb) in stand-off jamming configuration; maximum take-off 29,483kg (65,000lb)

Dimensions: span 16.15m (53ft); length 18.24m (59ft 10in); height 4.95m (16ft 3in); wing area 49.1m² (528.9sq ft)

Country of Origin: USA
Type: shipboard variable-geometry multi-role fighter
Accommodation: pilot and systems officer seated in tandem on Martin-Baker GRU7A rocket-assisted ejector seats
Armament (fixed): one General Electric M61A1 Vulcan rotary-barrel 20mm cannon in the port side of the forward fuselage
Armament (disposable): this is carried on four underfuselage points and on two hardpoints under the inner portions of the wings, up to a maximum weight of 6,577kg (14,500lb)
Electronics and operational equipment: communication and navigation equipment, plus Hughes AN/AWG-9 weapon control system with radar able to detect targets at ranges in excess of 315km (195 miles), track 24 simultaneously and attack six at varying altitudes simultaneously; Northrop TCS (Television Camera Set) for long-range identification

of targets; Kaiser AN/AVG-12 head-up display; Goodyear AN/ALE-39 chaff dispenser; and various electronic countermeasures systems

Powerplant and fuel system: two 9,480kg (20,900lb) afterburning thrust Pratt & Whitney TF30-P-412A turbofans

Performance: maximum speed 2,486km/h (1,545mph) or Mach 2.34 at high altitude; cruising speed 1,019km/h (633mph); service ceiling more than 15,240m (50,000ft); range about 3,219km (2,000 miles) with maximum internal and external fuel

Weights: empty 18,036kg (39,762lb); normal take-off 26,553kg (58,539lb); maximum take-off 33,724kg (74,348lb)

Dimensions: span 19.54m (64ft 1.5in) spread and 11.65m (38ft 2.5in) swept; length 19.10m (62ft 8in); height 4.88m (16ft); wing area 52.49m² (565sq ft) spread

Grumman (General Dynamics) EF-111A Electric Fox

Country of Origin: USA
Type: tactical electronic countermeasures aircraft
Accommodation: pilot and electronics officer/navigator seated side-by-side in a McDonnell Douglas escape module with 18,144kg (40,000lb) thrust rocket motor
Armament (fixed): none
Armament (disposable): none
Electronics and operational equipment: communication and navigation equipment, plus IBM 4 Pi digital computer, Texas Instruments AN/APQ-110 terrain-following radar, AN/APQ-160 attack radar, Eaton AN/ALQ-99E tactical jamming system, Sanders AN/ALQ-137(V)3 electronic countermeasures self-protection system, AN/ALR-62(V)4 terminal threat-warning system, AN/ALR-23 radar countermeasures receiver system and AN/ALE-28 electronic countermeasures dispenser system; the possibility is being examined of fitting Westinghouse AN/ALQ-131 jammer pods under the wings

Powerplant and fuel system: two 8,392kg (18,500lb) afterburning thrust Pratt & Whitney TF30-P-3 turbofans, and a total internal fuel capacity of 18,919 litres (4,162 Imp gal) in wing and fuselage tanks

Performance: maximum speed 2,216km/h (1,377mph) or Mach 2.09 at high altitude; cruising speed 595km/h (370mph) for stand-off role, 856km/h (532mph) for close-support role, and 940km/h (584mph) for penetration role; initial climb rate 1,006m (3,300ft) at intermediate power; service ceiling 13,715m (45,000ft); range 1,497km (930-mile) radius for penetration mission, or 3,706km (2,303 miles) for ferrying

Weights: empty 25,072kg (55,275lb); normal loaded 31,751kg (70,000lb); maximum take-off 40,370kg (89,000lb)

Dimensions: span 19.2m (63ft) spread and 9.74m (31ft 11.4in) swept; length 23.16m (76ft); height 6.1m (20ft); wing area 48.77m² (525sq ft) spread

OV-1D Mohawk

Grumman OV-1B Mohawk

Country of Origin: USA
Type: multi-sensor observation aircraft
Accommodation: pilot and systems operator seated
side-by-side on ejector seats

Armament (fixed): none

Armament (disposable): provision for bombs, rocket-launchers or 7.62mm (.3in) Minigun pods on two underwing hardpoints

Electronics and operational equipment: communication and navigation equipment, plus AN/AAS-4 infra-red reconnaissance or AN/APS-94 side-looking airborne radar (SLAR) systems, two KA-60C 180° cameras and one KA-76 serial-frame camera

Powerplant and fuel system: two 1,044kW (1,400shp) Avco Lycoming T53-L-701 turboprops

Performance: maximum speed 465km/h (289mph) on a SLAR mission at 3,050m (10,000ft) service ceiling 7,620m (25,000ft); range 1,520km (944 miles)

Weights: empty 5,468kg (12,054lb); normal take-off 7,140kg (15,741lb) for SLAR missions

Dimensions: span 14.63m (48ft); length 12.50m (41ft) excluding SLAR pod; height 3.86m (12ft 8in); wing area 33.44m² (360sq ft)

117

Hindustan Aeronautics Ltd Ajeet

Country of Origin: India
Type: lightweight interceptor and attack aircraft
Accommodation: pilot seated on a Martin-Baker GF4 lightweight ejector seat
Armament (fixed): two Aden Mk 4 30mm cannon with 90 rounds per gun
Armament (disposable): this is carried on four underwing hardpoints, the inner pair each capable of accepting a 227kg (500lb) bomb, or a BTV cluster bomb, or a BL-755 cluster bomb, or a CBLS-200/IA cluster bomb, or a pod for Russian 57mm (2.24in) rockets, or a Type 122 pod for 18 68mm (2.68in) rockets, and the outer pair each capable of taking a Type 122 pod, or a Russian pod, or a 136 litre (30 Imp gal) drop-tank
Electronics and operational equipment: communication and navigation equipment, plus a Ferranti F 195R/3 ISIS weapon sight

Powerplant and fuel system: one 2,041kg (4,500lb) thrust Rolls-Royce Orpheus Mk 701-01 turbojet, and a total internal fuel capacity of 1,350 litres (297 Imp gal) in nine fuselage and two integral wing tanks, plus provision for 273 litres (60 Imp gal) of external fuel

Performance: maximum speed 1,022km/h (635mph) or Mach .96 at 11,890m (39,000ft), and 1,102km/h (685mph) or Mach .9 at sea level; climb to 11,890m (39,000ft) in 6 minutes 2 seconds from brakes-off; service ceiling 13,715m (45,000ft); range 172km (107-mile) radius on a low-level attack mission with two 227kg (500lb) bombs

Weights: empty 2,307kg (5,085lb); normal take-off 3,538kg (7,800lb); maximum take-off 4,173kg (9,200lb)

Dimensions: span 6.73m (22ft 1in); length 9.04m (29ft 8in); height 2.46m (8ft 1in); wing area 12.69m² (136.6sq ft)

Hindustan Aeronautics Ltd HF-24 Marut Mk I

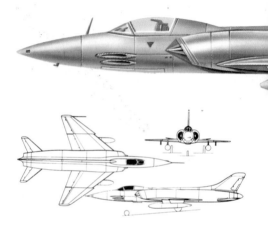

Country of Origin: India
Type: ground-attack fighter
Accommodation: pilot seated on a Martin-Baker Mk S4C ejector seat
Armament (fixed): four Aden Mk 2 30mm cannon in the nose with 120 rounds per gun, and a retractable Matra 103 pack of 50 68mm (2.68in) rockets in the lower fuselage

Armament (disposable): up to a maximum weight of 1,814kg (4,000lb)

Electronics and operational equipment: communication and navigation equipment, plus a Ferranti ISIS gyro gunsight

Powerplant and fuel system: two 2,200kg (4,850lb) thrust HAL-built Rolls-Royce (Bristol) Orpheus Mk 703 turbojets

Performance: maximum speed 1,086km/h (675mph) or Mach 1.02 at 12,190m (40,000ft); climb to 12,190m (40,000ft) in 9.33 minutes; range (HF-24 Marut Mk IT) 1,445km (898-mile) high-altitude interception radius

Weights: empty 6,195kg (13,658lb); normal take-off 6,951kg (19,734kg); maximum take-off 10,925kg (24,085lb)

Dimensions: span 9m (29ft 6.25in); length 15.87m (52ft 0.75in); height 3.60m (11ft 9.75in); wing area 28m² (301.4sq ft)

Ilyushin Il-28 'Beagle'

Country of Origin: USSR
Type: light bomber
Accommodation: pilot, bombardier/navigator and rear gunner in separate positions

Armament: two NR-23 23mm cannon with 100 rounds per gun in the lower nose, and two NR-23 23mm cannon with 225 rounds per gun on a flexible mounting in the rear turret; the maximum bombload is 3,000kg (6,614lb)

Powerplant and fuel system: two 2,700kg (5,952lb) thrust Klimov VK-1A turbojets

Performance: maximum speed 900km/h (559mph) at 4,500m (14,765ft) and 800km (497mph) at sea level; service ceiling 12,300m (40,355ft); range 2,180km (1,350 miles) with a 1,000kg (2,205lb) bombload

Weights: empty 13,000kg (26,455lb); normal take-off 18,400kg (40,565lb); maximum take-off 21,200kg (46,737lb)

Dimensions: span 21.45m (70ft 4.5in); length 17.65m (57ft 11in); height 6.7m (22ft); wing area 60.8m² (654.4sq ft)

Ilyushin Il-38 'May'

Country of Origin: USSR

Type: maritime reconnaissance and anti-submarine aircraft
Accommodation: (estimated) crew of three or four on the flightdeck, and a mission crew of eight or nine in the cabin
Armament (fixed): none

Armament (disposable): weapons such as depth charges, homing torpedoes and perhaps missiles are carried in a large bay in the lower fuselage; details of weapon types and total weapon load are not available

Electronics and operational equipment: communications and navigation equipment, plus search radar in an undernose radome, magnetic anomaly detection (MAD) gear in a tail 'sting', sonobuoys and onboard computing and analysis equipment

Powerplant and fuel system: four 3,169-kW (4,250-ehp) Ivchenko AI-20M turboprops

Performance: maximum speed 645km/h (401mph); cruising speed 645km/h (401mph) at 8,230m (27,000ft); range 7,240km (4,500 miles)

Weights: empty about 38,000kg (83,775lb); maximum take-off about 61,000kg (134,480lb)

Dimensions: span 37.4m (122ft 8.5in); length 39.6m (129ft 10in); height 10.17m (33ft 4in); wing area 140m² (1,507sq ft)

Ilyushin Il-76M 'Candid'

Country of Origin: USSR
Type: heavy transport
Accommodation: crew of seven including two freight-handlers for a payload of up to 40,000kg (88,183lb); alternatively, up to 140 troops can be accommodated

Armament (fixed): two NR-23 23mm cannon on a flexible mounting in the rear turret
Armament (disposable): none
Electronics and operational equipment: communi-

603679

cation and navigation equipment, plus weather/ navigation radar, computer-controlled automatic light-control system, and sophisticated freight-handling system

Powerplant and fuel system: four 12,000kg (26,455lb) thrust Soloviev D-30KP turbofans

Performance: maximum speed 850km/h (528mph); cruising speed 800km/h (497mph); service ceiling 15,500m (50,855ft); range 5,000km (3,107 miles) with maximum payload

Weights: empty about 62,000kg (136,684lb); maximum take-off 170,000kg (374,780lb)

Dimensions: span 50.5m (165ft 8in); length 46.59m (152ft 10.5in); height 14.76m (48ft 5in); wing area 300m² (3,229.2sq ft)

127

Israel Aircraft Industries Kfir-C2

Country of Origin: Israel
Type: interceptor and ground-attack aircraft
Accommodation: pilot seated on a Martin-Baker IL10P ejector seat

Armament (fixed): two DEFA 552 30mm cannon with 140 rounds per gun
Armament (disposable): up to a maximum weight of 5,775kg (12,731lb)
Electronics and operational equipment: communication and navigation equipment, plus MBT ASW-41 control-augmentation system, MBT ASW-42 stability-augmentation system, Elbit S-8600 (licence-built Singer-Kearfott) multi-mode navigation system and Rafael Mahat weapon-delivery system (or IAI WDNS 141 weapon-delivery and navigation system), Tamam central air-data computer, Elta EL/M-2001B or EL/M-2021 pulse-Doppler air-to-air and air-to-surface target-acquisition and target-tracking radar, Israel Electro-Optics head-up display and automatic gunsight, and ECM equipment (internal and podded)

Powerplant and fuel system: one 8,119kg (17,900lb) afterburning thrust General Electric J79-J1E turbojet

Performance: maximum speed over 2,440km/h (1,516mph) or Mach 2.3 at 11,000m (36,090ft), and 1,390km/h (864mph) or Mach 1.1 at sea level; service ceiling 17,680m (58,000ft) in stable flight in air-combat configuration; range 345km (214-mile)

Weights: empty 7,285kg (16,060lb); normal take-off 9,390kg (20,701lb) for interception or 14,670kg (32,341lb) for ground-attack; maximum take-off 16,200kg (35,714lb)

Dimensions: span 8.22m (26ft 11.5in) for wing, and 3.73m (12ft 3in) for canard; length 15.65m (51ft 4.5in) including probe; wing area 34.8m² (374.6sq ft)

Lockheed F-104G Starfighter

Lockheed F 104S Starfighter
fitted with Sparrow and Sidewinder
air/air missiles

Country of Origin: USA
Type: multi-role fighter
Accommodation: pilot seated on a Martin-Baker Mk
GQ7(F) ejector seat

Armament (fixed): one General Electric M61A-1 Vulcan rotary-barrel 30mm cannon with 750 rounds in the nose

Armament (disposable): up to a maximum weight of 1,955kg (4,310lb)

Electronics and operational equipment: communication and navigation equipment, plus Autonetics F15A NASARR fire-control radar system, General Electric AN/ASG-14 optical sight system, Honeywell MH-97 automatic flight-control system and an inertial navigation system

Powerplant and fuel system: one 7,167kg (15,800lb) afterburning thrust General Electric J79-GE-11A or MAN/Turbo-Union J79-MTU-J1K

Performance: maximum speed 2,333km/h (1,450mph) at 10,970m (36,000ft); service ceiling 17,680m (58,000ft); range 2,495km (1,550 miles)

Weights: 6,758kg (14,900lb); normal take-off 9,838kg (21,690lb)

Dimensions: span 6.68m (21ft 11in); length 16.69m (54ft 9in); wing area 18.22m^2 (196.1sq ft)

Lockheed P-2H Neptune

Country of Origin: USA
Type: long-range anti-submarine and maritime patrol aircraft

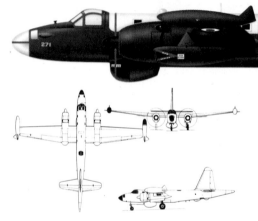

Accommodation: crew of three on the flightdeck, and four systems operators
Armament (fixed): provision for two 12.7mm (.5in) Colt-Browning M2 machine-guns in a dorsal turret
Armament (disposable): up to 3,629kg (8,000lb) of depth charges, torpedoes and bombs

Electronics and operational equipment: communication and navigation equipment, plus AN/APS-20 search radar with its antenna in a ventral radome, magnetic anomaly detection (MAD) gear with its sensor in a tail 'string', Julie active and Jezebel passive sonar systems and other specialist items

Powerplant and fuel system: two 2,610kW (3,500hp) Wright R-3350-32W radial piston engines and two 1,542kg (3,400lb) thrust Westinghouse J34-WE-34 turbojets

Performance: maximum speed 648km/h (403mph) at 6,095m (20,000ft); cruising speed 370km/h (230mph); service ceiling 9,145m (30,000ft); range 4,450km (2,765 miles) with internal fuel

Weights: empty 22,650kg (49,935lb); maximum take-off 36,240kg (79,895lb)

Dimensions: span 30.87m (101ft 3.5in); length 29.23m (95ft 10.75in); height 8.93m (29ft 3.5in); wing area 92.9m² (1,000sq ft)

Lockheed P-3C Orion

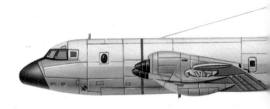

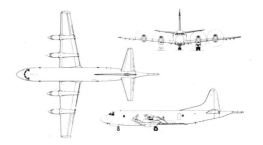

Country of Origin: USA
Type: long-range anti-submarine aircraft
Accommodation: flightcrew of five, and a mission crew of five in the cabin
Armament (fixed): none

Armament (disposable): up to a maximum weight of 9,070kg (20,000lb)

Electronics and operational equipment: communication and navigation equipment

Powerplant and fuel system: four 3,661-kW (4,910-ehp) Allison T56-A-14 turboprops

Performance: maximum speed 761km/h (473mph) at 4,570m (15,000ft); cruising speed 608km/h (378mph) at 7,620m (25,000ft); initial climb rate 594m (1,950ft) per minute; service ceiling 8,625m (28,300ft); range 2,494km (1,550-mile) radius with 3-hour patrol, or 3,835km (2,383-mile) radius with no time on station

Weights: empty 27,892kg (61,490lb); normal take-off 61,235kg (135,000lb); maximum take-off 64,410kg (142,000lb)

Dimensions: span 30.87m (99ft 8in); length 35.61m (116ft 10in); height 10.29m (33ft 8.5in); wing area 120.77m² (1,300sq ft)

135

Lockheed S-3A Viking

Country of Origin: USA
Type: shipboard anti-submarine aircraft
Accommodation: crew of two on the flightdeck, and tactical crew of two in the cabin, all seated on Douglas Escapac 1-E ejector seats
Armament (fixed): none

Armament (disposable): the weapons bay can accommodate four 454kg (1,000lb) destructors, or four Mk 46 torpedoes
Electronics and operational equipment: communication and navigation equipment, plus a tactical system centred on a Univac AN/AYK-10 digital computer receiving inputs from the Texas Instruments AN/APS-116 high-resolution search radar, OR-89/AA forward-looking infra-red (FLIR) in a retractable turret and other systems
Powerplant and fuel system: two 4,207kg (9,275lb) thrust General Electric TF34-GE-2 turbofans

Performance: maximum speed 834km/h (518mph); cruising speed 686km/h (426mph); initial climb rate more than 1,280m (4,200ft) per minute; service ceiling more than 10,670m (35,000ft); range more than 3,701km (2,300 miles) for combat, and more than 5,560km (3,455 miles) for ferrying

Weights: empty 12,088kg (26,650lb); normal take-off 19,277kg (42,500lb); maximum take-off 23,832kg (52,540lb)

Dimensions: span 20.93m (68ft 8in); length 16.26m (53ft 4in); height 6.93m (22ft 9in); wing area 55.56m² (598sq ft)

Lockheed SR-71A

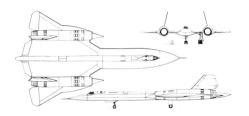

Country of Origin: USA
Type: strategic reconnaissance aircraft
Accommodation: pilot and reconnaissance-systems
operator seated in tandem on ejector seats

Armament (fixed): none

Armament (disposable): none

Electronics and operational equipment: communication and navigation equipment, plus a number of classified reconnaissance sensors

Powerplant and fuel system: two 14,742kg (32,500lb) afterburning thrust Pratt & Whitney JT11D-20B bleed turbojets (turbo-ramjets), and a total internal fuel capacity of more than 45,460 litres (10,000 Imp gal) in integral tanks; inflight-refuelling capability

Performance: maximum speed 3,620km/h (2,250mph) or Mach 3.4 at high altitude; cruising speed 3,186km/h (1,980mph) or Mach 3 at high altitude; service ceiling 30,480m (100,000ft); range 4,800km (2,980 miles) on internal fuel

Weights: maximum take-off 77,111kg (170,000lb)

Dimensions: span 16.94m (55ft 7in); length 32.74m (107ft 5in); height 5.64m (18ft 6in)

McDonnell Douglas A-4M Skyhawk II

Country of Origin: USA
Type: carrierborne light attack bomber
Accommodation: pilot seated on a Douglas Escapac 1-G3 lightweight ejector seat
Armament (fixed): two Mk 12 20mm cannon with 200 rounds per gun in the wing roots
Armament (disposable): this is carried on one under-fuselage hardpoint, rated at 1,588kg (3,500lb), and on four underwing hardpoints, the inner pair each rated at 1,021kg (2,250lb) and the outer pair each at 454kg (1,000lb); a great variety of weapon loads can be carried, including nuclear bombs, the Mk 84 907kg (2,000lb) bomb, the Mk 83 454kg (1,000lb) free-fall or retarded bomb, the Mk 82 227kg (500lb) free-fall or retarded bomb, the Mk 81 113kg (250lb) free-fall or retarded bomb, the LAU-3/A launcher with 19 69.85mm (2.75in) rockets, the LAU-10/A launcher with four 127mm (5in) rockets

Electronics and operational equipment: communication and navigation equipment, plus Bendix automatic flight control, Marconi AN/AVQ-24 head-up display, Texas Instruments AN/AJB-3 bombing system, AN/ASN-41 navigation computer, AN/APN-153(V) radar navigation, and electronic countermeasures

Powerplant and fuel system: one 5,080kg (11,200lb) Pratt & Whitney J52-P-408 turbojet

Performance: maximum speed 646km/h (1,040mph) with 1,814kg (4,000lb) bombload; initial climb rate 3,140m (10,300ft) per minute; range 3,220km (2,000 miles) with maximum fuel

Weights: empty 4,899kg (10,800lb); maximum take-off 11,113kg (24,500lb)

Dimensions: span 8.38m (27ft 6in); length 12.29m (40ft 4in) excluding probe; height 4.57m (15ft 10in); wing area 24.16m² (260sq ft)

McDonnell Douglas/British Aerospace AV-8B Harrier II

Country of Origin: USA/UK
Type: V/STOL close-support aircraft
Accommodation: pilot seated on a Stencel ejector seat (US aircraft)
Armament (fixed): one General Electric GAU-12/A 25mm rotary-barrel cannon system

Armament (disposable): to a maximum of 3,175kg (7,000lb)

Electronics and operational equipment: communication and navigation equipment, plus a Garrett digital air-data computer, inertial navigation system, 360° radar-warning receiver

Powerplant and fuel system: one 9,979kg (22,000lb) vectored-thrust Rolls-Royce Pegasus 11-21 (F402-RR-406) turbofan

Performance: maximum speed 988km/h (615mph) or Mach .93 at 10,975m (36,000ft); range 282km (172 mile) radius with 12 Mk 82 Snakeye bombs

Weights: empty 5,783kg (12,750lb); normal take-off 10,410kg (22,950lb); maximum take-off 13,494kg (29,750lb)

Dimensions: span 9.25m (30ft 4in); length 14.12m (46ft 4in); height 3.55m (11ft 7.75in); wing area 21.37m² (230sq ft)

McDonnell Douglas CF-101B Voodoo

Country of Origin: USA
Type: long-range interceptor fighter
Accommodation: pilot and weapons officer
Armament (fixed): none
Armament (disposable): comprises three Hughes AIM-4D Falcon air-to-air missiles in the weapons bay and two Douglas Air-2A Genie

Electronics and operational equipment: communication and navigation equipment, plus MG-13 fire-control system with automatic search and tracking modes for the radar

Powerplant and fuel system: two 7,666kg (16,900lb) afterburning thrust Pratt & Whitney J57-P-55 turbojets, and a total internal fuel capacity of 7,771 litres (1,709 Imp gal) in the centre fuselage and wings

Performance: maximum speed 1,835km/h (1,134mph) or Mach 1.7 at 10,670m (35,000ft); cruising speed 887km/h (551mph); initial climb rate 914,995m (49,200ft) per minute; service ceiling 16,705m (54,800ft); range 2,445km (1,520 miles)

Weights: empty 13,141kg (28,970lb); normal take-off 20,713kg (45,664lb); maximum take-off 23,768kg (52,400lb)

Dimensions: span 12.09m (39ft 8in); length 20.55m (67ft 5in); height 5.49m (18ft); wing area 34.19m² (368sq ft)

McDonnell Douglas F/A-18A Hornet

Country of Origin: USA
Type: shipboard strike fighter
Accommodation: pilot seated on a Martin-Baker US10S ejector (perhaps to be replaced by a Stencel seat)
Armament: one General Electric M61A1 Vulcan rotary-barrel 20mm cannon in the nose; maximum load is 7,711kg (17,000lb)
Electronics and operational equipment: communication and navigation equipment, plus Hughes AN/APG-65 multi-mode air-to-air and air-to-surface

tracking radar, Kaiser head-up display, Itek AN/ALR-67 radar-warning receiver, Litton inertial navigation system, General Electric flight-control system with two AN/AYK-14 digital computers, and provision for a Martin Marietta AN/ASQ-173 laser spot tracker and Ford FLIR

Powerplant and fuel system: two 7,257kg (16,000lb) afterburning thrust General Electric F404-GE-400 turbofans

Performance: maximum speed more than 1,915km/h (1,190mph) or Mach 1.8 at high altitude; service ceiling 15,240m (50,000ft); range more than 740km (460-mile) fighter mission radius, or 1,019km (633-mile) attack-mission radius, or 3,700km (2,300 miles) for ferrying

Weights: empty 9,331kg (20,570lb); normal take-off 15,234kg (33,585lb) for a fighter mission; maximum take-off 21,887kg (48,252lb) for an attack mission

Dimensions: span 11.43m (36ft 6in) without missiles; length 17.07m (56ft); height 4.66m (15ft 3.5in); wing area 37.16m² (400sq ft)

147

McDonnell Douglas F-4E Phantom II

Country of Origin: USA

Type: all-weather multi-role fighter

Accommodation: two seated in tandem on Martin-Baker Mk H7 ejector seats

Armament (fixed): one 20mm General Electric M61A-Vulcan rotary-barrel cannon with 640 rounds

Armament (disposable): this is accommodated on four recessed stations under the fuselage (each capable of accepting an AIM-7 Sparrow AAM) and five hardpoints (one under the ventral fuselage and four under the wings); the fuselage hardpoint can accept nuclear stores (B28, B43, B57 or B61 bombs) up to a weight of 986kg (2,170lb) or up to 1,371kg (3,020lb) of conventional stores; four wing points can accommodate up to 5,888kg (12,980lb)

Electronics and operational equipment: AN/APN-155 radar altimeter; AN/AJB-7 all-altitude bombing system; AN/ASN-46A navigation computer; AN/ASN-63 inertial navigation system; AN/ASQ-91 weapon-release system; AN/ASG-26 lead-comput-

ing optical sight; AN/APR-36 radar homing and warning receiver; AN/ASA-32 automatic flight control system; AN/APQ-120 radar fire control system; and CPK-92A computer

Powerplant and fuel system: two 8,119kg (17,900lb) thrust General Electric J79-GE-17A afterburning turbojets

Performance: maximum speed (clean) 2,301km/h (1,430mph) or Mach 2.17 at 10,975m (36,000ft); cruising speed 917km/h (570mph) with stores; initial climb rate (clean) 15,180m (49,800ft) per minute; service ceiling (clean) 17,905m (58,750ft); range 1,145km (712-mile) combat radius on an interdiction mission, and 3,184km (1,978 miles) for ferrying

Weights: empty 13,757kg (30,328lb); normal take-off 18,818kg (41,487lb); maximum take-off 28,030kg (61,795lb)

Dimensions: span 11.77m (38ft 7.5in); length 19.20m (63ft); height 5.02m (16ft 5.5in); wing area 49.24m^2 (530sq ft)

McDonnell Douglas F-15C Eagle

Country of Origin: USA
Type: air-superiority and attack fighter
Accommodation: pilot seated on a Douglas ACES II ejector seat
Armament (fixed): one General Electric M61A1 Vulcan rotary-barrel 20mm cannon with 940 rounds in the upper edge of the starboard inlet
Armament (disposable): this is carried on special positions (AIM-7 Sparrow air-to-air missiles) and on three underfuselage and two underwing hardpoints, up to a maximum of 7,257kg (16,000lb); the missile load is normally four AIM-7F Sparrow and four AIM-9L Sidewinder medium and short-range air-to-air missiles; other stores can include the Mk 84 907kg (2,000lb) free-fall or guided and other weapons
Electronics and operational equipment: communication and navigation equipment, plus Hughes AN/APG-63 pulse-Doppler search and tracking radar, McDonnell Douglas head-up display, IBM central

computer, Northrop AN/ALQ-135 internal electronic countermeasures system, AN/ALR-56 radar-warning receiver, Magnavox electronic warfare warning system and other items

Powerplant and fuel system: two 10,864kg (23,950lb) afterburning thrust Pratt & Whitney F100-PW-100 turbofans

Performance: maximum speed more than 2,655km/h (1,650mph) or Mach 2.5 at high altitude; absolute ceiling 30,480m (100,000ft); range more than 4,631km (2,878 miles) without FAST packs but with drop-tanks, or more than 5,560km (3,450 miles) with maximum fuel

Weights: empty 12,700kg (28,000lb); normal take-off 20,212kg (44,560lb) for interception mission; maximum take-off 30,845kg (68,000lb)

Dimensions: span 13.05m (42ft 9.75in); length 19.43m (63ft 9in); height 5.63m (18ft 5.5in); wing area 56.5m² (608sq ft)

Mikoyan-Gurevich MiG-15bis 'Fagot'

Country of Origin: USSR
Type: fighter
Accommodation: pilot seated
on an ejector seat

MiG-15UTI 'Midget' two-seat
operational trainer version

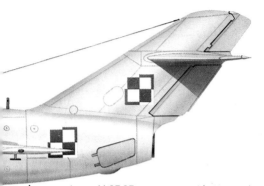

Armament: one N-37 37mm cannon with 40 rounds and two NR-23 23mm cannon with 80 rounds per gun in the nose; maximum weight of 1,000kg (2,205lb) of bombs or rockets

Electronics and operational equipment: communication and navigation equipment

Powerplant and fuel system: one 3,170kg (6,989lb) thrust Klimov VK-1A turbojet

Performance: maximum speed 1,100km/h (684mph) at 12,000m (39,370ft); service ceiling 15,550m (51,015ft); range 2,000km (1,232 miles) with maximum fuel

Weights: empty 3,400kg (7,495lb); normal take-off 4,960kg (10,934lb)

Dimensions: span 10.08m (33ft.75in); length 11.05m (36ft3.25in); height 3.4m (11ft1.75in); wing area 20.6m² (221.7sq ft)

153

Mikoyan-Gurevich MiG-17F 'Fresco-C'

Country of Origin: USSR
Type: fighter
Accommodation: pilot seated on an ejector seat

Armament (fixed): one N-37D 37mm cannon with 40 rounds and two NR-23 23mm cannon with 80 rounds

Electronics and operational equipment: communication and navigation equipment
Powerplant and fuel system: one 3,400kg (7,495lb) afterburning thrust Klimov VK-1F turbojet
Performance: maximum speed 1,145km/h (711mph) or Mach .97 at 3,000m (9,840ft); initial climb rate 3,900m (12,795ft) per minute; service ceiling 16,600m (54,460ft); range 1,470km (913 miles)
Weights: empty 4,100kg (9,040lb); normal take-off 5,340kg (11,773lb); maximum take-off 6,700kg (14,770lb)
Dimensions: span 9.63m (31ft 7in); length 11.26m (36ft 11.25in); height 3.35m (11ft); wing area 22.6m² (243.3sq ft)

155

Mikoyan-Gurevich MiG-19SF 'Farmer-C'

Country of Origin: USSR
Type: fighter
Accommodation: pilot seated on an ejector seat

Armament (fixed): three NR-30 30mm cannon
Armament (disposable): this is carried on two underwing hardpoints, and can consist of two AA-2 'Atoll' air-to-air missiles, or two 212mm (8.35in) rockets, or two packs each with eight 57mm (2.24in) rockets, or two 250 or 500kg (551 or 1,102lb) bombs

Electronics and operational equipment: communication and navigation equipment

Powerplant and fuel system: two 3,250kg (7,165lb) afterburning thrust Klimov RD-9BF turbojets

Performance: maximum speed 1,450km/h (901mph) or Mach 1.35 at 10,000m (32,810ft); cruising speed 950km/h (590mph) at 10,000m (32,810ft); initial climb rate 6,900m (22,635ft) per minute; service ceiling 17,900m (58,725ft); range 1,390km (863 miles) on internal fuel

Weights: empty 5,170kg (11,397lb); normal take-off 7,400kg (16,314lb); maximum take-off 8,900kg (19,621lb)

Dimensions: span 9.2m (30ft 2.25in); length 12.6m (41ft 4in) excluding probe; height 3.9ft (12ft 9.5in); wing area 25m² (269.1sq ft)

Mikoyan-Gurevich MiG-21MF 'Fishbed-J'

The 1971 version of the MiG-21 MF is the
MiG-21 SMT 'Fishbed-K with improved
fuel capacity

Country of Origin: USSR

Type: multi-role fighter

Accommodation: pilot seated on ejector seat

Armament (fixed): one twin-barrel GSh-23 23mm
cannon with 200 rounds in a belly pack

Armament (disposable): this is carried on four
underwing hardpoints, up to a maximum weight of
about 1,500kg (3,307lb)

Electronics and operational equipment: communication and navigation equipment, plus 'Jay Bird' search and tracking radar in inlet centrebody and a gyro gunsight

Powerplant and fuel system: one 6,600kg (14,550lb) Tumansky R-13-300 turbojet

Performance: maximum speed 2,230km/h (1,385mph) or Mach 2.1 at 11,000m (36,090ft) and 1,300km/h (807mph) or Mach 1.06 at sea level; service ceiling about 15,250m (50,030ft); range 370km (230-mile) hi-lo-hi radius with four 250kg (551lb) bombs, or 1,100km (683 miles) 'clean' on internal fuel, or 1,800km (1,118 miles) for ferrying with three drop-tanks

Weights: normal take-off 8,200kg (18,077lb) with four AA-2 missiles; maximum take-off 9,400kg (20,723lb) with two AA-2 missiles and three drop-tanks

Dimensions: span 7.15m (23ft 5.5in); length 15.76m (51ft 8.5in) including probe; height 4.10m (13ft 5.5in); wing area 23m² (247.6sq ft)

Mikoyan-Gurevich MiG-23MF 'Flogger-G'

The first version in service was the
MiG-23S all weather interceptor

Country of Origin: USSR

Type: variable-geometry air-combat fighter

Accommodation: pilot only, seated on an ejector seat

Armament: one GSh-23 23mm twin-barrel cannon in a fuselage belly pack; maximum of about 2,000kg (4,409lb) of stores

Electronics and operational equipment: communication and navigation equipment, plus 'High Lark'

search radar (with a search range of 85km/53 miles and a tracking range of 54km/34 miles), undernose laser ranger, Sirena 3 radar-warning system, Doppler navigation and ECM gear

Powerplant and fuel system: one 12,475kg (27,502lb) afterburning thrust Tumansky R-29 turbo-jet, and a total internal fuel capacity of 5,750 litres (1,265 Imp gal), plus provision for one 800 litre (176 Imp gal) drop-tank on the centreline hardpoint

Performance: maximum speed 2,500km/h (1,553mph), or Mach 2.35 at high altitude and 1,470km/h (913mph) or Mach 1.2 at sea level; service ceiling 18,600m (61,025ft); combat radius between 900 and 1,200km (560 and 745 miles)

Weights: normal take-off 12,700kg (27,998lb); maximum take-off 16,000kg (35,273lb)

Dimensions: span spread 14.25m (46ft 9in) and swept 8.17m (26ft 9.5in); length 16.8m (55ft 1.5in); height 4.35m (14ft 4in); wing area 37m² (398.3sq ft)

Mikoyan-Gurevich MiG-25 'Foxbat-A'

Country of Origin: USSR
Type: interceptor fighter
Accommodation: pilot only, seated on a KM-1 ejector seat
Armament (fixed): none
Armament (disposable): this is carried on four underwing hardpoints, and generally comprises four AA-6 'Acrid' air-to-air missiles, or two AA-7 'Apex' and two AA-8 'Aphid' air-to-air missiles

Electronics and operational equipment: communication and navigation equipment, plus 'Fox Fire' radar (with a range of 85km/52 miles and look-down capability) in the nose and Sirena 3 radar-warning receiver

Powerplant and fuel system: two 11,000kg (24,250lb) afterburning thrust Tumansky R-31 turbojets, and a total internal fuel capacity of about 17,410 litres (3,830 Imp gal) in fuselage, inlet saddle and integral wing tanks

Performance: maximum speed 2,975km/h (1,849mph) or Mach 2.8 at high altitude with four AA-6 'Acrid' air-to-air missiles, and 1,040km/h (646mph) at sea level with four AA-6 'Acrid' air-to-air' missiles; initial climb rate 12,480m (40,945ft) per minute; service ceiling 24,400m (80,050ft); combat radius 1,130km (702 miles)

Weights: empty about 20,000kg (44,092lb); maximum take-off 36,200kg (79,806lb)

Dimensions: span 13.95m (45ft 9in); length 23.82m (78ft 1.75in); height 6.1m (20ft .25in); wing area 56.83m² (611.7sq ft)

Mikoyan-Gurevich MiG-27 'Flogger-D'

Country of Origin: USSR
Type: ground-attack aircraft
Accommodation: pilot only

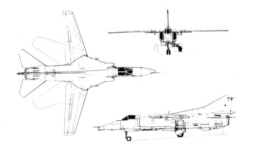

Armament: one 23mm six-barrel rotary cannon in a ventral package plus maximum of 3,000kg (6,614lb) of stores

Electronics and operational equipment: communication and navigation equipment, plus a laser ranger and marked-target seeker and Sirena 3 radar-warning receiver

Powerplant and fuel system: one 11,500kg (25,353lb) afterburning thrust Tumansky R-29B

Performance: maximum speed 1,595km/h (991mph) or Mach 1.5 at high altitude; service ceiling 16,000m (52,495ft); ferry range with three drop-tanks 2,500km (1,553 miles)

Weights: normal take-off 15,500kg (34,170lb); maximum take-off 18,000kg (39,863lb)

Dimensions: span spread 14.25m (46ft 9in) and swept 8.17m (26ft 9.5in); length 16m (52ft 6in); height 4.35m (14ft 4in); wing area spread 27.26m² (293.4sq ft)

Mitsubishi F-1

Country of Origin: Japan
Type: close-support fighter
Accommodation: pilot seated on a Daiseru-built Weber ES-7J ejector seat
Armament (fixed): one JM61A-1 Vulcan rotary-barrel 20mm cannon
Armament (disposable): this is carried on one under-fuselage and four underwing hardpoints, up to a maximum weight of 2,722kg (6,000lb)
Electronics and operational equipment: communication and navigation equipment, plus Mitsubishi

Electric air-to-air and air-to-surface radar, Mitsubishi Electric (Thomson-CSF) J/AWG-11 head-up display, Ferranti 6TNJ-F inertial navigation system, Mitsubishi Electric J/AWG-12 fire-control and bombing computation system, and a radar warning receiver

Powerplant and fuel system: two 3,207kg (7,070lb) afterburning thrust Rolls-Royce/Turboméca Adour Mk 801A turbofans

Performance: maximum speed 1,700km/h (1,056mph) or Mach 1.6 at 10,970m (36,000ft); initial climb rate 10,670m (35,000ft) per minute; service ceiling 15,240m (50,000ft); range 555km (345-mile) hi-lo-hi radius with two ASM-1s and one drop-tank or 2,600km (1,616 miles) with maximum internal and external fuel

Weights: empty 6,358kg (14,017lb); normal take-off 9,860kg (21,737lb); maximum take-off 13,675kg (30,148lb)

Dimensions: span 7.88m (25ft 10.25in); length 17.84m (58ft 6.25in) including probe; height 4.28m (14ft 4.25in); wing area 21.18m² (288sq ft)

North American F-86F Sabre

Country of Origin: USA
Type: fighter and fighter-bomber
Accommodation: pilot seated on an ejector seat
Armament (fixed): six 12.7mm (.5in) Colt-Browning M3 machne-guns in the nose with 267 rounds per gun
Armament (disposable): provision under the wings for two AIM-9 Sidewinder air-to-air missiles, or two 454kg (1,000lb) bombs, or eight rockets
Electronics and operational equipment: communication and navigation equipment, plus ranging radar in the nose

Powerplant and fuel system: one 2,708kg (5,970lb) thrust General Electric J47-GE-27 turbojet.

Performance: maximum speed 1,105km/h (687mph) at sea level; initial climb rate 3,050m (10,000ft) per minute; service ceiling 15,240m (50,000ft); range 1,485km (925 miles) on internal fuel, and 2,044km (1,250 miles) with drop-tanks

Weights: empty 5,045kg (11,125lb); normal take-off 7,711kg (17,000lb); maximum take-off 9,350kg (20,610lb)

Dimensions: span 11.91m (39ft 1in); length 11.44m (37ft 6.5in); height 4.47m (14ft 8.75in); wing area 26.76m^2 (288sq ft)

North American F-100D Super Sabre

Country of Origin: USA
Type: interceptor and fighter-bomber
Accommodation: pilot seated on an ejector seat

Armament (fixed): four M39 20mm cannon with
200 rounds per gun in the fuselage
Armament (disposable): up to a maximum weight of
3,402kg (7,500lb)

Electronics and operational equipment: communication and navigation equipment, plus attack radar

Powerplant and fuel system: one 7,711kg (17,000lb) afterburning thrust Pratt & Whitney J57-P-21A turbojet

Performance: maximum speed 1,392km/h (865mph) or Mach 1.31 at 10,670m (35,000ft); cruising speed 909km/h (565mph) at 10,970m (36,000ft) initial climb rate 4,875m (16,000ft) per minute; service ceiling 13,715m (45,000ft); range 853km (530-mile) radius

Weights: empty 9,525kg (21,000lb); maximum take-off 15,800kg (34,830lb)

Dimensions: span 11.81m (38ft 9in); length 16.54m (54ft 3in) including probe; height 4.96m (16ft 2.75in); wing area 35.77m² (385sq ft)

Northrop F-5A Freedom Fighter

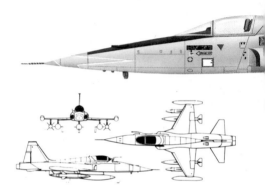

Country of Origin: USA
Type: lightweight tactical fighter
Accommodation: pilot seated on a rocket-assisted ejector seat
Armament (fixed): two Colt-Browning M39 20mm cannon in the nose with 280 rounds per gun
Armament (disposable): this is carried on one under-fuselage and four underwing hardpoints and on two wingtip missile-launcher rails, up to a maximum weight of 1,996kg (4,400lb)

172

Electronics and operational equipment: communication and navigation equipment, plus a Norsight optical sight and control equipment for the AGM-12 Bullpup missile when appropriate

Powerplant and fuel system: two 1,850kg (4,080lb) afterburning thrust General Electric J85-GE-13

Performance: maximum speed 1,489km/h (925mph) or Mach 1.4 at 10,970m (36,000ft); cruising speed 1,031km/h (640mph) or Mach .97 at 10,670m (36,000ft); initial climb rate 8,750m (28,700ft) per minute; service ceiling 15,390m (50,500ft); range 314km (195-mile) radius with maximum payload, or 2,594km (1,612 miles) with maximum internal and external fuel

Weights: empty 3,667kg (8,085lb); maximum take-off 9,379kg (20,677lb)

Dimensions: span 7.7m (25ft 3in); length 14.38m (47ft 2in); height 4.01m (13ft 2in); wing area 15.79m² (170sq ft)

173

Northrop F-5E Tiger II

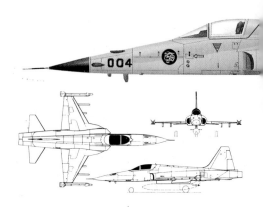

Country of Origin: USA
Type: lightweight tactical fighter
Accommodation: pilot seated on a rocket-assisted ejector seat
Armament (fixed): two Colt-Browning M39A2 20mm cannon with 280 rounds per gun in the nose
Armament (disposable): this is carried on one under-fuselage and four underwing hardpoints and on two wingtip missile rails, up to a maximum of 3,175kg (7,000lb)

Electronics and operational equipment: communication and navigation equipment, plus AN/APQ-159 lightweight air-to-air search and tracking radar, and AN/ASG-29 optical gunsight

Powerplant and fuel system: two 2,267kg (5,000lb) afterburning thrust General Electric J85-GE-21

Performance: maximum speed 1,730km/h (1,075mph) or Mach 1.63 at 10,970m (36,000ft); cruising speed 1,038km/h (645mph) or Mach .98 at 10,970m (36,000ft); initial climb rate 9,630m (31,600ft) per minute; service ceiling 15,850m (52,000ft); range 305km (190-mile) radius with two Sidewinders and 2,857kg (6,300lb) dropload, or 3,720km (2,314 miles) with maximum fuel

Weights: empty 4,275kg (9,425lb); maximum take-off 11,561kg (25,488lb)

Dimensions: span 8.13m (26ft 8in); length 14.73m (48ft 3.75in); height 4.08m (13ft 4.5in); wing area 17.3m² (186sq ft)

Northrop F-20 Tigershark

Country of Origin: USA
Type: lightweight tactical fighter
Accommodation: pilot seated on a rocket-assisted ejector seat
Armament (fixed): two Colt-Browning M39A2 20mm cannon with 450 rounds per gun in the nose
Armament (disposable): this is carried on one under-fuselage and four underwing hardpoints, and on two wingtip missile rails, to a maximum weight of more than 3,629kg (8,000lb); typical offensive loads are three General Electric GPU-5/A 30mm cannon pods, or six AIM-9 Sidewinder air-to-air missiles, or four AGM-65 Maverick air-to-surface missiles, or nine Mk 82 227kg (500lb) free-fall or retarded bombs, or four guided bombs; and among the other weapons which can be carried are the Mk 84 907kg (2,000lb) bomb, the Mk 83 454kg (1,000lb) bomb, BLU-series napalm bombs, CBU-series bomb dispensers, LAU-series rocket-launcher pods
Electronics and operational equipment: communication and navigation equipment, plus General Elec-

tric GE-200 multi-mode radar with look-up/look-down capability, General Electric head-up display, Teledyne digital central computer, Honeywell inertial navigation system, AN/ALR-46 radar-warning receiver, AN/ALE-40 countermeasures dispenser system, and AN/ALQ-171(V) conformal countermeasures system

Powerplant and fuel system: one 7,711kg (17,000lb) afterburning thrust General Electric F404-GE-100 turbofan

Performance: maximum speed about 2,125km/h (1,320mph) or Mach 2 at high altitude; initial climb rate 16,490m (54,100ft) per minute; service ceiling 16,765m (55,000ft); range 556km (345-mile) combat-air-patrol radius

Weights: empty 5,089kg (11,220lb); normal take-off 6,831kg (15,060lb) with half fuel; maximum take-off 11,295kg (26,290lb)

Dimensions: span 8.13m (26ft 8in); length 14.17m (46ft 6in) excluding probe; height 4.22m (13ft 10.25in)

Panavia Tornado IDS

Country of Origin: Italy/UK/West Germany
Type: variable-geometry multi-role combat aircraft
Accommodation: pilot and systems operator seated in tandem on Martin-Baker Mk 10A ejector seats

Armament (fixed): two IWKA-Mauser 27mm cannon with 360 rounds per gun
Armament (disposable): up to a weight of about 8,165kg (18,000lb)
Electronics and operational equipment: communication and navigation equipment, plus Texas Instruments multi-mode forward-looking radar, Ferranti FIN 1010 digital inertial navigation/radar display system, Decca Type 72 Doppler radar navigation, Microtecnica air-data computer, Litef Spirit 3 central digital computer, Smiths/Teldix/OMI head-up display, Ferranti laser range and marked-target seeker, Elettronica radar-warning receiver, Marconi/Plessey/Decca Sky Shadow active ECM equipment, and Marconi/Selenia stores-management system

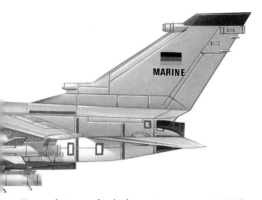

Powerplant and fuel system: two 7,620kg (16,800lb) afterburning thrust Turbo-Union RB.199-34R Mk 103 turbofans

Performance: maximum speed more than 2,125km/h (1,320mph) or Mach 2 at 11,000m (36,090ft); service ceiling more than 15,000m (49,210ft); range 1,390km (863 miles)

Weights: empty 14,090kg (31,063lb); normal take-off 20,410kg (44,996lb) with maximum internal fuel but no stores; maximum take-off 26,490kg (58,399lb)

Dimensions: span 13.9m (45ft 7.25in) spread, and 8.60m (28ft 2.5in) swept; length 16.7m (54ft 9.5in); height 5.7m (18ft 8.5in); wing area about 25m² (269sq ft)

Rockwell B-1B

Country of Origin: USA
Type: variable-geometry long-range strategic bomber and missile-carrier
Accommodation: crew of four on the flightdeck, all seated on Douglas ACES ejector seats
Armament (fixed): none
Armament (disposable): this is carried in three lower-fuselage weapons bays, up to a maximum weight of 29,030kg (64,000lb), and on eight under-fuselage hardpoints, up to a maximum weight of 12,701kg (28,000lb); in the strategic nuclear role, the weapons bays can accommodate 24 B-61 or B-83 bombs, or 12 B-28 or B-43 bombs, or eight AGM-86B air-launched cruise missiles, or 24 AGM-69 short-range attack missiles, while the underfuselage hardpoints can accept 14 B-43, B-61 or B-83 bombs, or eight B-28 bombs, or 14 AGM-86B air-launched cruise missiles or AGM-69 short-range attack missiles; in the conventional role, the weapons bays can lift 24 Mk 84 907kg (2,000lb)

Electronics and operational equipment: communication and navigation equipment, plus Westinghouse Offensive Radar System based on the AN/APG-66 multi-mode radar, Eaton (AIL) Defensive Avionics System based on the AN/ALQ-161 electronic counter-measure system, and Boeing Offensive Avionics System including the Offensive Radar System plus Singer-Kearfott inertial navigation and AN/APN-218 Doppler navigation

Powerplant and fuel system: four 13,608kg (30,000lb) afterburning thrust General Electric F101-GE-102 turbofans

Performance: maximum speed about Mach 1.25, or 965km/h (600mph) or Mach .79 for penetration at 61m (200ft); range about 12,000km (7,455 miles)

Weights: maximum take-off 216,365kg (477,000lb)

Dimensions: span 41.67m (136ft 8.5in) spread and 23.84m (78ft 25in) swept; length 44.81m (147ft); height 10.36m (34ft); wing area about 181.2m² (1,950sq ft)

Rockwell OV-10A Bronco

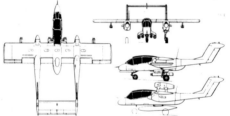

Country of Origin: USA
Type: multi-role counter-insurgency aircraft
Accommodation: crew of two seated in tandem on
LW-3B ejector seats
Armament (fixed): two 7.62mm (.3in) machine-guns
with 500 rounds per gun in each sponson
Armament (disposable): up to a maximum weight of
1,633kg (3,600lb)

Electronics and operational equipment: communication and navigation equipment

Powerplant and fuel system: two 533kW (715ehp) Garrett T76-G-416-417 turboprops, and a total internal fuel capacity of 954 litres (210 Imp gal) in five wing tanks, plus provision for one 568-litre (125-Imp gal) drop-tank

Performance: maximum speed 452km/h (281mph) at sea level; initial climb rate 790m (2,600ft) per minute; service ceiling 7,315m (24,000ft); range 367km (228-mile) radius with maximum payload, or 2,224km (1,382 miles) for ferrying with maximum fuel

Weights: empty 3,127kg (6,893lb); normal take-off 4,494kg (9,908lb); maximum take-off 6,552kg (14,444lb)

Dimensions: span 12.19m (40ft); length 12.67m (41ft 7in); height 4.62m (15ft 2in); wing area 27.03m² (291sq ft)

Rockwell T-39A

Country of Origin: USA
Type: administrative support and pilot-proficiency aircraft

Accommodation: crew of two on the flightdeck, and up to six passengers in the cabin
Armament (fixed): none

Armament (disposable): none

Electronics and operational equipment: communication and navigation equipment

Powerplant and fuel system: two 1,361kg (3,000lb) thrust Pratt & Whitney J60-P-3 turbojets, and a total internal fuel capacity of 4,024 litres (885 Imp gal) in one fuselage and integral wing tanks

Performance: maximum speed 958km/h (595mph) at 10,970m (36,000ft); cruising speed 727km/h (452mph) at 12,190m (40,000ft); initial climb rate 1,695m (5,550ft) per minute; service ceiling 12,190m (40,000ft); range 2,776km (1,725 miles)

Weights: empty 4,218kg (9,300lb); maximum take-off 8,056kg (17,760lb)

Dimensions: span 13.54m (45ft 5.25in); length 13.34m (43ft 9in); height 4.88m (16ft); wing area 31.78m² (342.05sq ft)

Saab 35X Draken

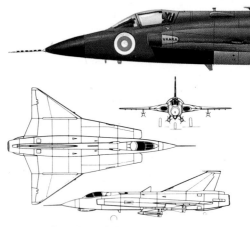

Country of Origin: Sweden
Type: all-weather fighter, attack and reconnaissance aircraft
Accommodation: pilot seated on a Saab 73SE-F rocket-assisted ejector seat
Armament: two Aden 30mm cannon in the wings with 100 rounds per gun, plus up to 4,500kg (9,921lb) stores

Electronics and operational equipment: communication and navigation equipment, plus Saab S7BX weapon-delivery radar and computer system and a Saab BT9 toss-bombing computer

Powerplant and fuel system: one 8,000kg (17,650lb) afterburning thrust Volvo Flygmotor RM6C turbojet (licence-built version of the Rolls-Royce Avon Series 300)

Performance: maximum speed 2,125km/h (1,320mph) or Mach 2 at high altitude; initial climb rate 10,500m (34,450ft) per minute; service ceiling about 19,800m (64,960ft); range 635km (395-mile) hi-lo-hi radius, or 3,250km (2,020 miles) for ferrying with maximum internal and external fuel

Weights: normal take-off 11,400kg (25,130lb); maximum take-off 16,000kg (35,275lb)

Dimensions: span 9.4m (30ft 10in); length 15.35m (50ft 4in); height 3.89m (12ft 9in); wing area 49.20m² (529.6sq ft)

Saab 91B Safir

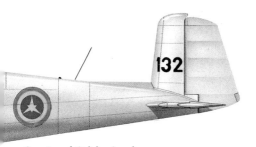

Country of Origin: Sweden
Type: basic training and liaison aircraft
Accommodation: pupil and instructor seated side-by-side at the front of the cockpit, and up to two passengers seated at the rear of the cockpit
Armament (fixed): none
Armament (disposable): none
Electronics and operational equipment: communication and navigation equipment
Powerplant and fuel system: one 142kW (190hp) Avco Lycoming O-435-A flat-six piston engine
Performance: maximum speed 275km/h (171mph); cruising speed 244km/h (152mph); initial climb rate 348m (1,142ft) per minute; service ceiling 6,250m (20,505ft); range 1,075km (668 miles)
Weights: empty 720kg (1,587lb); maximum take-off 1,220kg (2,690lb)
Dimensions: span 10.6m (34ft 9.25in); length 7.92m (25ft 11.75in); height 2.2m (7ft 2.5in); wing area 13.6m² (146.39sq ft)

Saab 105

The Saab 105G is fitted with a naval attack system

Country of Origin: Sweden
Type: trainer and light attack aircraft
Accommodation: pupil and instructor seated side-by-side on ejector seats
Armament (fixed): none

Armament (disposable): up to a maximum weight of 2,000kg (4,409lb) stores

Electronics and operational equipment: communication and navigation equipment

Powerplant and fuel system: two 1,293kg (2,850lb) thrust General Electric J85-GE-17B turbojets, and a total internal fuel capacity of 1,400 litres (310 Imp gal) in two fuselage and two integral wing tanks; provision for drop-tanks

Performance: maximum speed 950km/h (603mph) at sea level; climb to 10,000m (32,810ft) in 4.5 minutes; service ceiling 13,000m (42,650ft); range 2,400km (1,491 miles)

Weights: empty 2,550kg (5,534lb); maximum take-off 6,500kg (14,330lb)

Dimensions: span 9.50m (31ft 2in); length 10.5m (34ft 5in); height 2.70m (8ft 10in); wing area 16.30m² (175.46sq ft)

Saab-Scania AJ 37 Viggen

Country of Origin: Sweden
Type: all-weather attack aircraft
Accommodation: pilot seated on a Saab-Scania
rocket-assisted ejector seat

Armament (fixed): none
Armament (disposable): up to a maximum weight of
6,000kg (13,228lb)
Electronics and operational equipment: communi-
cation and navigation equipment, plus L.M. Ericsson
UAP-1023 search and attack radar, Saab-Scania CK-
37 central computer, Phillips air-data computer,
Marconi head-up display, Decca Type 72 Doppler
navigation, SATT radar-warning receiver, Svenska
Radio radar display and electronic countermeasures
systems
Powerplant and fuel system: one 11,800kg
(26,015lb) afterburning thrust Volvo Flygmotor
RM8A (licence-built and modified Pratt & Whitney
JT8D-22) turbofan

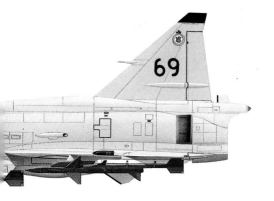

Performance: maximum speed 2,125km/h (1,320mph) or Mach 2 at 12,000m (39,370ft) and more than 1,335km/h (830mph) or Mach 1.1 at sea level; climb to 10,000m (32,810ft) in less than 1 minute 40 seconds; service ceiling about 15,200m (49,870ft); range at least 1,000km (621-mile) hi-lo-hi radius with external weapons, and 500km (311-mile) lo-lo-lo radius with external weapons

Weights: empty about 11,800kg (26,015lb); normal take-off 15,000kg (33,069lb); maximum take-off 20,500kg (45,194lb)

Dimensions: span 10.60m (34ft 9.25in), and canard 5.45m (17ft 10.5in); length 16.30m (53ft 5.75in) including probe; height 5.80m (19ft .25in); wing area 46m² (495.1sq ft) and canard 6.2m² (66.74sq ft)

SEPECAT Jaguar A and S

Country of Origin: France/UK
Type: close support and reconnaissance aircraft
Accommodation: pilot seated on a Martin-Baker 9B Mk II ejector seat
Armament (fixed): two Aden 30mm cannon with 150 rounds per gun
Armament (disposable): this is carried on one under-fuselage hardpoint (rated at 1,134kg/2,500lb) and two underwing hardpoints (the inner pair each rated at 1,134kg/2,500lb and the outer pair each rated at 567kg/1,250lb), up to a maximum weight of 4,763kg/10,500lb
Electronics and operational equipment: communication and navigation equipment, plus (French aircraft) CSF 121 fire-control unit, CSF 21 weapon-aiming computer, Dassault fire-control computer for Martel operation and CSF laser rangefinder; or (British aircraft) Marconi NAVWASS navigation and weapon-aiming system (being replaced by Ferranti

FIN 1064 digital inertial navigation and weapon-aiming system), Smiths head-up display and Ferranti laser range and marked-target seeker

Powerplant and fuel system: two 3,647kg (8,040lb) afterburning thrust Rolls-Royce/Turboméca Adour Mk 104 turbofans

Performance: maximum speed 1,700km/h (1,056mph) or Mach 1.6 at 11,000m (36,090ft), and 1,350km/h (840mph) or Mach 1.1 at sea-level; climb to 9,145m (30,000ft) in 1 minute 30 seconds; service ceiling 14,000m (45,930ft); range 537km (334-mile) lo-lo-lo radius with internal fuel, or 3,525km (2,190 miles) for ferrying with external fuel

Weights: empty 7,000kg (15,432lb); normal take-off 10,955kg (24,150lb); maximum take-off 15,700kg (34,612lb)

Dimensions: span 8.69m (28ft 6in); length 16.83m (55ft 2.5in) including probe; height 4.89m (16ft .5in); wing area 24.18m^2 (260.27sq ft)

Shenyang J-6C

Country of Origin: China
Type: fighter, attack and reconnaissance aircraft
Accommodation: pilot on a Martin-Baker PKD10 ejector seat
Armament (fixed): two or three NR-30 30mm cannon
Armament (disposable): this is accommodated on four underwing hardpoints, the outer pair normally being used for drop-tanks; typical loads are four air-to-air missiles, or eight 212mm (8.35in) rockets, or two 250kg (551lb) bombs
Electronics and operational equipment: communication and navigation equipment
Powerplant and fuel system: two 3,250kg (7,165lb) afterburning thrust Wopen-6 (Tumansky R-9BF-811) turbojets, and a total internal fuel capacity of 2,170

litres (477 Imp gal) in two main and two smaller
fuselage tanks, plus provision for two 760 or 1,140-
litre (167 or 251-Imp gal) drop-tanks

Powerplant and fuel system: maximum speed
1,540km/h (957mph) or Mach 1.45 at 11,000m
(36,090ft) and 1,340km/h (832mph) or Mach 1.09 at
sea level; cruising speed 950km/h (590mph); initial
climb rate more than 9,145m (30,000ft) per minute;
service ceiling 17,900m (58,725ft); range 685km
(426-mile) radius with two 760-litre (167-Imp gal)
drop-tanks and 2,200km (1,366 miles) maximum

Performance: empty 5,670kg (12,700lb); normal
take-off 7,545kg (16,634lb); maximum take-off
8,965kg (19,764lb)

Weights: span 9.20m (30ft 2.25in); length 12.6m
(41ft 4in) excluding nose probe; height 3.88m (12ft
8.75in); wing area 25m² (269.1sq ft)

Shin Meiwa PS-1

Country of Origin: Japan
Type: STOL anti-submarine flying-boat
Accommodation: crew of five on the flightdeck, and a mission crew of five in the cabin
Armament (fixed): none
Armament (disposable): this is carried in an internal weapons bay, two underwing pods and two wingtip hardpoints; the weapons bay can carry four 150kg (331lb) anti-submarine bombs, the pods can each accommodate two homing torpedoes
Electronics and operational equipment: communication and navigation equipment, plus AN/APS-80 search radar with its antenna in the thimble nose, AN/HQS-101B dunking sonar, AN/ASQ-10A magnetic anomaly detection (MAD) gear, AN/AQA-3 Jezebel passive sonar with 20 sonobuoys, Julie active sonar ranging equipment, AN/APN-153 Doppler

radar, AN/AYK-2 navigation computer, N-OA-35/
HSA tactical plotting group and other items

Powerplant and fuel system: four 2,283kW (3,060-
ehp) Ishikawajima-built General Electric T64-IHI-10
turboprops

Performance: maximum speed 547km/h (340mph)
at 1,525m (5,000ft); cruising speed 315km/h
(196mph) at 1,525m (5,000ft) on two engines; initial
climb rate 690m (2,264ft) per minute; service ceiling
9,000m (29,530ft); range 2,170km (1,348 miles) for
a normal mission, or 4,745km (2,948 miles) for
ferrying

Weights: empty 26,300kg (58,000lb); normal take-
off 36,000kg (79,365lb); maximum take-off
43,000kg (94,797lb)

Dimensions: span 33.14m (108ft 8.75in); length
33.50m (109ft 11in); height 9.715m (31ft 10.5in);
wing area 135.8m² (1,461sq ft)

SIAI-Marchetti SF.260TP

Country of Origin: Italy
Type: trainer and tactical support aircraft
Accommodation: pilot and co-pilot seated side-by-side, plus provision for a passenger to the rear of the cockpit

Armament (fixed): none

Armament (disposable): up to a maximum weight of 300kg (661lb)

Electronics and operational equipment: communication and navigation equipment

Powerplant and fuel system: one Allison 250-B17C turboprop flat-rated to 261kW (350shp)

Performance: maximum speed 380km/h (236mph) at sea level; service ceiling 8,535m (28,000ft); range 950km (590 miles) at 4,570m (15,000ft)

Weights: empty 795kg (1,753lb); maximum take-off 1,200kg (2,646lb) as a trainer, and 1,300kg (2,866lb) as an armed aircraft

Dimensions: span 8.35m (27ft 4.75in) over tiptanks; length 7.40m (24ft 3.25in); height 2.41m (7ft 11in); wing area 10.1m² (108.7sq ft)

Soko J-1 Jastreb

Country of Origin: Yugoslavia
Type: light attack aircraft

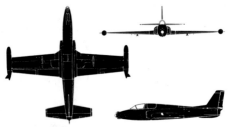

Accommodation: pilot seated on an HSA (Folland) Type 1-B lightweight ejector seat
Armament (fixed): three 12.7mm (.5in) Colt-Browning machine-guns with 135 rounds per gun in the nose
Armament (disposable): this is carried on eight underwing hardpoints; the inner pair are able to

accept two 250kg (551lb) bombs, or two clusters of smaller bombs, or two 150-litre (33-Imp gal) napalm tanks, or two launchers each with 12 57mm (2.24in) rockets; and the outer six are each able to lift one 127mm (5in) rocket

Electronics and operational equipment: communication and navigation equipment

Powerplant and fuel system: one 1,361kg (3,000lb) thrust Rolls-Royce (Bristol) Viper Mk 531 turbojet

Performance: maximum speed 820km/h (510mph) at 6,000m (19,685ft); cruising speed 740km/h (460mph) at 5,000m (16,405ft); initial climb rate 1,260m (4,135ft) per minute; service ceiling 12,000m (39,375ft); range 1,520km (945 miles) with maximum fuel

Weights: empty 2,820kg (6,217lb); maximum take-off 5,100kg (11,243lb)

Dimensions: span 11.68m (38ft 4in) over tiptanks; length 10.88m (35ft 8.5in); height 3.64m (11ft 11.5in); wing area 19.43m² (209.14sq ft)

Sukhoi Su-7BMK 'Fitter-A'

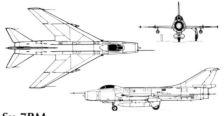

Su-7BM

Country of Origin: USSR
Type: ground-attack fighter
Accommodation: pilot seated on an ejector seat
Armament: two NR-30 30mm cannon with 70 rounds per gun in the wing roots, plus up to a nominal weight of 2,500kg (5,511lb) of stores

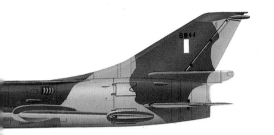

Electronics and operational equipment: communication and navigation equipment, plus ranging radar in the inlet centrebody, Sirena 3 radar-warning receiver and ASP-5PF gyro sight

Powerplant and fuel system: one 10,000kg (22,046lb) afterburning thrust Lyulka AL-7F-1

Performance: maximum speed 1,700km/h (1,055mph) or Mach 1.6 at 11,000m (36,090ft), and 1,350km/h (840mph) or Mach 1.1 at sea level; initial climb rate about 9,120m (29,920ft) per minute; service ceiling 15,150m (49,700ft); range 345km (215-mile) combat radius with two drop-tanks, or 1,450km (901 miles) for ferrying with maximum fuel

Weights: empty 8,620kg (19,004lb), normal take-off 12,000kg (26,455lb); maximum take-off 13,500kg (29,762lb)

Dimensions: span 8.93m (29ft 3.5in); length 17.37m (57ft) including probe; height 4.57m (15ft); wing area 31.5m² (339.1sq ft)

Sukhoi Su-11 'Fishpot-C'

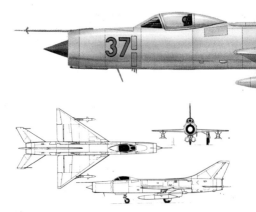

Country of Origin: USSR
Type: all-weather fighter
Accommodation: pilot seated on an ejector seat
Armament (fixed): none
Armament (disposable): this is carried on two underwing hardpoints, and comprises one AA-3 'Anab' IR-homing and one AA-3-2 'Improved Anab' radar-homing air-to-air missile
Electronics and operational equipment: communication and navigation equipment, plus Uragan 5B

'Skip Spin' interception radar and Sirena 3 radar-warning receiver

Powerplant and fuel system: one 10,000kg (22,046lb) afterburning thrust Lyulka Al-7F-1 turbojet, and a total internal fuel capacity of 4,000 litres (880 Imp gal) in fuselage and integral wing tanks, plus provision for two 600-litre (132-Imp gal) drop-tanks on hardpoints under the fuselage

Performance: maximum speed 1,915km/h (1,190mph) or Mach 1.8 at 11,000m (36,090ft) without drop-tanks, or 1,160km/h (720mph) or Mach .95 at 300m (985ft); initial climb rate 8,200m (26,905ft) per minute; service ceiling 17,000m (55,775ft); range about 1,125km (700 miles)

Weights: empty about 8,400kg (18,519lb); maximum take-off about 13,600kg (29,980lb)

Dimensions: span 8.43m (27ft 8in); length 17m (56ft) including probe; height 4.88n (16ft); wing area about 28.0m² (301.4sq ft)

Sukhoi Su-15 'Flagon-F'

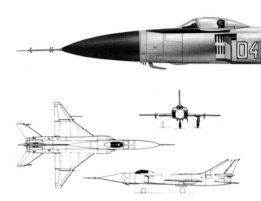

Country of Origin: USSR
Type: all-weather interceptor fighter
Accommodation: pilot seated on ejector seat
Armament (fixed): possibly one GSh-23 23mm twin-barrel cannon in the lower fuselage
Armament (disposable): this is carried on four underwing hardpoints, and generally comprises two IR-homing AA-3 'Anab' and two radar-homing AA-3-2 'Advanced Anab' air-to-air missiles, the underfuselage hardpoints are normally used for the carriage of drop-tanks, but are believed to be capable of accepting weapons

208

Electronics and operational equipment: communication and navigation equipment, plus 'Skip Spin' interception radar in the inlet centrebody, Sirena 3 radar-warning receiver in the vertical tail and other systems

Powerplant and fuel system: probably two 7,200kg (15,873lb) afterburning thrust Tumansky R-13F2-300 turbojets

Performance: maximum speed about 2,655km/h (1,650mph) or Mach 2.5 at high altitude without stores, or 2,370km/h (1,473mph) or Mach 2.3 with stores, climb to 11,000m (36,090ft) in 2.5 minutes; service ceiling 20,000m (65,615ft); range 725km (450-mile) combat radius

Weights: empty 12,500kg (27,557lb); normal take-off 16,000kg (35,273lb); maximum take-off 20,500kg (45,194lb)

Dimensions: span 10.53m (34ft 6in); length 20.5m (68ft); height 5m (16ft 5in); wing area 36m² (387.7sq ft)

Sukhoi Su-17 'Fitter-C'

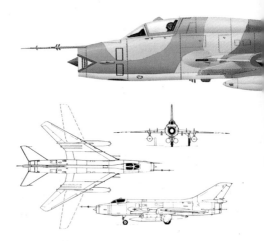

Country of Origin: USSR
Type: variable-geometry ground-attack fighter
Accommodation: pilot seated on an ejector seat
Armament: two NR-30 30mm cannon with 70 rounds per gun in the wing roots; bombload maximum weight of 4,000kg (8,818lb)

Powerplant and fuel system: one 11,200kg (24,691lb) afterburning thrust Lyulka AL-21F-3

Performance: maximum speed 2,300km/h (1,429mph) or Mach 2.17 at high altitude, and 1,285km/h (798mph) or Mach 1.05 at sea level; initial climb rate 13,800m (45,275ft) per minute; service ceiling 18,000m (59,055ft); range 630 km (391-mile) hi-lo-hi radius with a 2,000kg (4,409lb) load, or 360km (224-mile) lo-lo-lo radius with the same load

Weights: empty 10,000kg (22,046lb); normal take-off 14,000kg (30,865lb); maximum take-off 17,700kg (39,020lb)

Dimensions: span 14m (45ft 11.25in) spread and 10.6m (34ft 9.5in) swept; length 18.75m (61ft 6.25in) including probes; height 4.75m (15ft 7in); wing area 40.1m^2 (431.6sq ft) spread and 37.2m^2 (400.4sq ft) swept

Sukhoi Su-24 'Fencer-C'

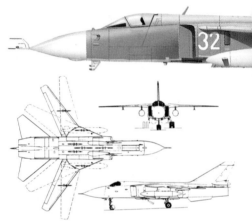

Country of Origin: USSR
Type: variable-geometry attack and interdiction aircraft
Accommodation: pilot and weapons officer seated side-by-side on ejector seats
Armament: possibly one GSh-23 23mm twin-barrel cannon under the port side of the fuselage, plus maximum load of 8,000kg (17,635lb)

Electronics and operational equipment: communication and navigation equipment, plus attack and navigation radars, radar-warning receiver and other systems including terrain-avoidance radar and (probably) laser marked-target seeker

Powerplant and fuel system: probably two 11,200kg (24,691lb) afterburning thrust Lyulka AL-21F-3

Performance: maximum speed more than 2,125km/h (1,320mph) or Mach 2 at high altitude; service ceiling 17,500m (57,415ft); range 1,800km (1,115-mile) hi-lo-hi radius with 2,000kg (4,409lb) payload and two external tanks, or 950km (590-mile) hi-lo-hi radius with 2,500kg (5,511lb) payload or 322km (200-mile) lo-lo-lo radius

Weights: maximum take-off 39,500kg (87,081lb)

Dimensions: span 17.15m (56ft 3in) spread and 9.53m (31ft 3in) swept, length 21.29m (69ft 10in); height 5.5m (18ft); wing area about 40m² (430.5sq ft)

213

Tupolev Tu-16 'Badger-A'

Country of Origin: USSR
Type: medium bomber and maritime reconnaissance aircraft

Accommodation: crew of six
Armament (fixed): two NR-23 23mm cannon each in dorsal and ventral barbettes, and in the radar-laid rear turret, plus one NR-23 23mm fixed cannon in the starboard side of the nose of aircraft without a radome

Armament (disposable): this is carried in a weapons bay in the lower part of the fuselage, up to a maximum of 9,000kg (19,841lb) of free-fall ordnance

Electronics and operational equipment: communication and navigation equipment, plus mapping radar and other systems

Powerplant and fuel system: two 9,500kg (20,944lb) thrust Mikulin AM-3M turbojets

Performance: maximum speed 990km/h (615mph) at 6,000m (19,685ft); cruising speed 850km/h (530mph) at optimum altitude; initial climb rate 1,250m (4,100ft) per minute; service ceiling 12,300m (40,355ft) range 2,900km (1,802-mile) radius without inflight-refuelling

Weights: empty 37,200kg (82,010lb); maximum take-off 72,000kg (158,730lb)

Dimensions: span 32.93in (108ft .5in); length 34.8m (114ft 2in); height 10.80m (35ft 6in); wing area 164.65m² (1,772.3sq ft)

Tupolev Tu-22 'Binder-A'

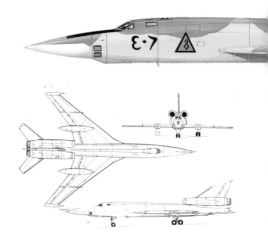

Country of Origin: USSR
Type: supersonic reconnaissance bomber
Accommodation: crew of three seated in tandem on ejector seats
Armament (fixed): one NS-23 23mm cannon in rear turret controlled by 'Bee Hind' gun-control radar

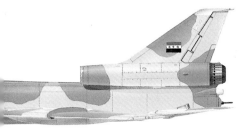

Armament (disposable): this is carried in a weapons bay in the lower fuselage, and comprises up to 10,000kg (22,046lb) of free-fall conventional or nuclear weapons

Electronics and operational equipment: communication and navigation equipment, plus attack radar and chaff dispensers

Powerplant and fuel system: two afterburning turbojets or unknown type, each rated at about 12,250kg (27,006lb) thrust, and a total internal capacity of about 45,500 litres (10,010 Imp gal) in fuselage and inner-wing tanks

Performance: maximum speed 1,480km/h (920mph) or Mach 1.4 at 12,200m (40,025ft); service ceiling 18,300m (60,040ft); range 3,100 km (1,926-mile) combat radius on internal fuel

Weights: empty 40,000kg (88,183lb); maximum take-off 83,900kg (184,965lb)

Dimensions: span 27.7m (90ft 10.5in); length 40.53m (132ft 11.5in); height 10.67m (35ft); wing area about 145m² (1,560.8sq ft)

Tupolev Tu-22M 'Backfire-B'

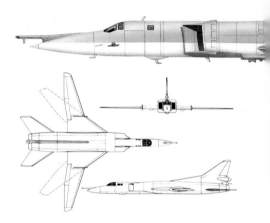

Country of Origin: USSR
Type: variable-geometry medium bomber and maritime reconnaissance aircraft
Accommodation: crew of four, probably seated on ejector seats
Armament: two 23mm cannon in a radar-controlled rear barbette with 'Bee Hind' gun-control radar, plus load of 12,000kg (26,455lb)

Electronics and operational equipment: communication and navigation equipment, plus 'Down Beat' attack and navigation radar in the nose

Powerplant and fuel system: two afterburning turbofans (possibly Kuznetsov NK-144 derivatives) each rated at about 20,000kg (44,092lb) thrust, and a total internal fuel capacity of about 60,000 litres (13,200 Imp gal) in fuselage and inner-wing tanks; inflight-refuelling capability

Performance: maximum speed 2,125km/h (1,320mph) or Mach 2 at high altitude, and 1,100km/h (684mph) or Mach .9 at sea level; service ceiling more than 16,000m (52,495f); range 5,470km (3,400-mile) radius on internal fuel

Weights: empty 50,000kg (110,230lb); maximum take-off 122,500kg (270,062lb)

Dimensions: span 34.45m (113ft) spread and 26.21m (86ft) swept; length 40.23m (132ft); height 10.06m (33ft); wing area about 165m² (1,776sq ft)

Tupolev Tu-28P 'Fiddler-B'

Country of Origin: USSR
Type: long-range all-weather interceptor

Accommodation: crew of two seated in tandem on ejector seats
Armament (fixed): none

Armament (disposable): this is carried on four underwing hardpoints, and comprises two IR-homing AA-5 'Ash' and two radar-homing AA-5 'Ash' air-to-air missiles

Electronics and operational equipment: communication and navigation equipment, plus 'Big Nose' attack radar

Powerplant and fuel system: two afterburning turbojets (possibly Lyulka AL-21F engines) each rated at about 12,250kg (27,006lb) thrust

Performance: maximum speed 1,850km/h (1,150mph) or Mach 1.75 at 11,000m (36,090ft); service ceiling 20,000m (65,615ft); range 4,990km (3,101 miles) with maximum fuel

Weights: empty 18,000kg (39,683lb); maximum take-off 45,000kg (99,206lb)

Dimensions: span 20m (65ft); length 26m (85ft); height 7m (23ft); wing area 80m² (861.1sq ft)

Tupolev Tu-95 'Bear-A'

Country of Origin: USSR
Type: long-range strategic heavy bomber

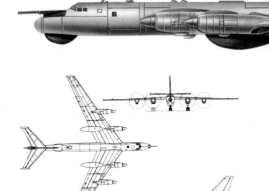

Accommodation: crew of 10
Armament (fixed): two NR-23 23mm cannon each
in remotely-controlled dorsal and ventral barbettes
and in the manned rear turret with 'Bee Hind' gun-
control radar

Armament (disposable): this is carried in a weapons bay in the lower fuselage, and comprises up to 11,340kg (25,000lb) of free-fall conventional or nuclear weapons

Electronics and operational equipment: communication and navigation equipment, plus attack radar with its antenna in a chin radome

Powerplant and fuel system: four 11,033kW (14,795ehp) Kuznetsov NK-12MV turboprops, and a total internal fuel capacity of 72,980 litres (16,540 Imp gal) in wing tanks

Performance: maximum speed 805km/h (500mph) at 12,500m (41,010ft); service ceiling about 13,400m (43,965ft); range 12,550km (7,798 miles) with maximum bombload, or a maximum radius of 8,285km (5,150 miles) on internal fuel

Weights: empty 75,000kg (165,344lb); normal take-off 150,000kg (330,688lb); maximum take-off 170,000kg (374,780lb)

Dimensions: span 48.5m (159ft); length 47.50m (155ft 10in); height 1,178m (38ft 8in)

Tupolev Tu- 'Blackjack'

Country of Origin: USSR
Type: variable-geometry strategic bomber
Accommodation: unknown
Armament (fixed): none
Armament (disposable): this is carried in weapons bays in the lower fuselage, and can comprise up to 16,000kg (35,273lb) of free-fall HE or nuclear bombs, though it is believed that long-range cruise missiles are under development as its longer-term primary armament
Electronics and operational equipment: communication and navigation equipment, plus (presumably) attack and navigation radars, radar-warning receiver and extensive ECM gear

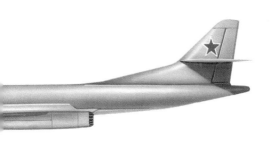

Powerplant and fuel system: four unidentified after-burning turbofans, possibly developments of the Kuznetsov NK-144 used in the unsuccessful Tu-144 SST, and each rated in the region of 22,500kg (49,603lb) thrust, and a total internal fuel capacity of about 165,000 litres (36,295 Imp gal); inflight-refuelling capability

Performance: maximum speed about 2,445km/h (1,519mph) or Mach 2.3 at high altitude; range about 13,500km (8,380 miles) on internal fuel

Weights: maximum take-off about 265,000kg (584,215lb)

Dimensions: span spread 42.7m (140ft 1in) and swept 29m (95ft 1.75in); length 54.85m (179ft 11.5in); height 13.7m (44ft 11.5in); wing area 237m² (2,551sq ft)

Tupolev Tu-126 'Moss'

Country of Origin: USSR
Type: airborne warning and control system aircraft
Accommodation: crew of 12
Armament (fixed): none
Armament (disposable): none

Electronics and operational equipment: communication and navigation equipment, plus search and tracking radar with its antenna in an 11m (36ft) rotodome above the rear fuselage, and several other systems including computerized data-processing, data-link communications and other aspects relating to the AWACS role

Powerplant and fuel system: four 11,033kW (14,795ehp) Kuznetov NK-12MV turboprops

Performance: maximum speed 850km/h (528mph) at high altitude; cruising speed 650km/h (404mph) at high altitude; service ceiling 10,000m (32,810ft); range 12,550km (7,800 miles) on internal fuel, sufficient for an endurance of more than 20 hours

Weights: empty about 90,000kg (194,415lb); maximum take-off 170,000kg (374,780lb)

Dimensions: span 51.2m (168ft); length 55.2m (181ft 1in); height 16.05in (52ft 8in); wing area 311.1m^2 (3,348.75sq ft)

Vought A-7E Corsair II

Country of Origin: USA
Type: shipboard attack aircraft and tactical fighter
Accommodation: pilot seated on a McDonnell Douglas Escapac ejector seat
Armament: one General Electric M61A1 Vulcan rotary-barrel 20mm cannon with 1,000 rounds in the port side of the fuselage, plus maximum stores of 6,804kg (15,000lbs)
Electronics and operational equipment: communication and navigation equipment, plus AN/APQ-126(V) multi-function radar, AN/ASN-91(V) navigation and weapon-delivery computer system, AN/ASN-90(V) inertial navigation, AN/APN-190(V) Doppler navigation, AN/AVQ-7(V) head-up display, CP-953A/AJQ air-data computer, AN/ASN-99 projected map display, AN/ALR-45/50 internal homing

and warning systems, and AN/ALQ-126 active electronic countermeasures

Powerplant and fuel system: one 6,804kg (15,000lb) thrust Allison TF41-A-2 (licence-built Rolls-Royce Spey) turbofan

Performance: maximum speed 1,110km/h (690mph) or Mach .9 at sea level, and 1,038km/h (645mph) or Mach .86 at 1,525km (5,000ft) with 12 Mk 82 227kg (500lb) bombs; range 3,669km (2,280 miles) on internal fuel, and 4,603 (2,860 miles) with maximum internal and external fuel

Weights: empty 8,676kg (19,127lb); maximum take-off 19,050kg (42,000lb)

Dimensions: span 11.8m (38ft 9in); length 14.06m (46ft 1.5in); height 4.9m (16ft .75in); wing area 34.83m² (375sq ft)

Vought F-8E Crusader

Country of Origin: USA
Type: shipboard fighter
Accommodation: pilot seated on an ejector seat
Armament (fixed): four Colt-Browning M39 20mm cannon with 144 rounds per gun in the forward fuselage

Armament (disposable): this is carried on the fuselage sides (four AIM-9 Sidewinder air-to-air missiles or eight 127mm (5in) rockets) and on two underwing hardpoints, up to a maximum weight of 2,268kg (5,000lb); typical underwing loads are two Mk 84 907kg (2,000lb) bombs, or two Mk 83 454kg (1,000lb) bombs, or four Mk82 227kg (500lb) bombs, or 12 Mk 81 113kg (250lb) bombs, or 24 127mm (5in) rockets; the French F-8E(FN) version carries two Matra R.530 air-to-air missiles on the sides of the fuselage
Electronics and operational equipment: communication and navigation equipment, plus AN/APQ-94 search and fire-control radar and other systems

Powerplant and fuel system: one 8,165kg (18,000lb) afterburning thrust Pratt & Whitney J57-P-20 turbojet

Performance: maximum speed 1,827km/h (1,135mph) or Mach 1.72 at 10,970m (36,000ft); cruising speed 901km/h (560mph) at 12,190m (40,000ft); initial climb rate about 6,400m (21,000ft) per minute; service ceiling 17,680m (58,000ft); range 966km (600-mile) combat radius

Weights: normal take-off 12,700kg (28,000lb); maximum take-off 15,420kg (34,000lb)

Dimensions: span 10.87m (35ft 8in); length 16.61m (54ft 6in); height 4.80m (15ft 9in); wing area 32.52m^2 (350sq ft)

Yakovlev Yak-28P 'Firebar'

Country of Origin: USSR
Type: all-weather fighter
Accommodation: crew of two seated in tandem on ejector seats
Armament (fixed): none

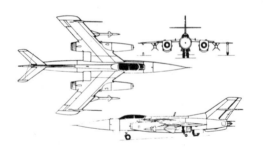

Armament (disposable): this is carried on two underwing hardpoints, and normally comprises one IR-homing AA-3 'Anab' and one radar-homing AA-3 'Anab' air-to-air missile
Electronics and operational equipment: communi-

cation and navigation equipment, plus attack radar in the nose and a radar-warning receiver in the tail, and other systems

Powerplant and fuel system: two 6,000kg (13,228lb) afterburning thrust turbojets (probably Tumansky R-11 units)

Performance: maximum speed 1,180km/h (733mph) or Mach 1.1 at 10,670m (35,000ft); cruising speed 920km/h (571mph) at optimum altitude, service ceiling 16,750m (55,000ft); range up to 2,575km (1,600 miles)

Weights: empty 13,600kg (29,982lb); maximum take-off 15,875kg (34,998lb)

Dimensions: span 12.95m (42ft 6in); length 23.17m (76ft); height 3.95m (12ft 11.5in)

Yakovlev Yak-36MP 'Forger-A'

Country of Origin: USSR
Type: shipboard VTOL combat aircraft
Accommodation: pilot seated on an ejector seat

Armament (fixed): none
Armament (disposable): this is carried on four underwing hardpoints, up to a maximum weight of about 1,360kg (3,000lb); typical stores are launchers for 57mm (2.24in) rockets, air-to-air missiles, air-to-surface missiles, bombs and gun pods containing the GSh-23 23mm twin-barrel cannon
Electronics and operational equipment: communication and navigation equipment, plus ranging radar

Powerplant and fuel system: one vectored-thrust turbojet in the rear fuselage, with a rating of about 7,950kg (17,526lb) thrust, and two direct-lift Kolesov turbojets in the forward fuselage, each with a rating of about 3,625kg (7,992lb) thrust

Performance: maximum speed 1,170km/h (725mph) or Mach 1.1 at high altitude, and 1,125km/h (700mph) or Mach .8 at sea level, initial climb rate 4,500m (14,765ft) per minute; service ceiling 12,000m (39,370ft), range 370km (230-mile) hi-lo-hi radius with maximum weapon load, or 240km (150-mile) lo-lo-lo radius with maximum weapon load

Weights: empty 7,485kg (16,500lb); maximum take-off 11,565kg (25,495lb)

Dimensions: span 7.32m (24ft); length 15.25m (50ft); height 4.37m (14ft 4in); wing area 16m² (172.22sq ft)

Index